CLIMATIC CATACLYSM

Kurt M. Campbell
EDITOR

CLIMATIC CATACLYSM

The Foreign Policy and National Security Implications of Climate Change

BROOKINGS INSTITUTION PRESS
Washington, D.C.

ABOUT BROOKINGS

THE BROOKINGS INSTITUTION
1775 Massachusetts Avenue, N.W., Washington, D.C. 20036
www.brookings.edu

Library of Congress Cataloging-in-Publication data

Climatic cataclysm : the foreign policy and national security implications of climate change / Kurt M. Campbell, editor.
p. cm.
Summary: "Presents three scenarios of what the future may hold: expected, severe, and catastrophic and analyzes the security implications of each. Considers what can be learned from early civilizations confronted with natural disaster and asks what the largest emitters of greenhouse gases can do to reduce and manage future risks"—Provided by publisher.
Includes bibliographical references and index.
ISBN 978-0-8157-1332-6 (cloth : alk. paper)
1. Climatic changes—Government policy. 2. Climatic changes—Forecasting. 3. Climatic changes—Social aspects. 4. International relations—Forecasting. 5. National security.
I. Campbell, Kurt M. II. Title.
QC981.8.C5C625 2008
363.738'74526—dc22 2008012194

9 8 7 6 5 4 3 2 1

The paper used in this publication meets minimum requirements of the American National Standard for Information Sciences—Permanence of Paper for Printed Library Materials: ANSI Z39.48-1992.

Typeset in Minion

Composition by Circle Graphics
Columbia, Maryland

Printed by R. R. Donnelley
Harrisonburg, Virginia

Contents

Acknowledgments

This project on climate change and national security fundamentally changed the perspective of its participants. Over the course of many months of meetings two very diverse groups—scientists and strategists—came together to explore the potentially profound implications of unchecked climate change on global security. The result was a sobering, sometimes even harrowing, set of assessments of what the world can expect if a carbon-based business-as-usual approach to civilization continues and expands in the years ahead. This volume is our attempt to concretely explore scenarios that most of us seem either to ignore or deny while we proceed with our daily tasks. The hope here is that by extrapolating on current trends and developments we might be better positioned to appreciate just how much current actions imperil future lives.

This book owes an enormous debt to several friends and colleagues who helped during the process of writing and researching. First, I would like to thank a distinguished group of nationally recognized leaders who helped to inform the discussion that took place over the course of a year, leading up to this book's publication. We identified and recruited these leaders from the fields of climate science, foreign policy, political science, oceanography, history, and national security to take part in this endeavor. Members of the group included Nobel Laureate Thomas Schelling; Pew Center Senior Scientist Jay Gulledge; National Academy of Sciences President Ralph Cicerone; American Meteorological Society Fellow Bob Correll; Woods Hole Oceanographic Institute Senior Scientist Terrence Joyce and former Vice President Richard Pittenger; Climate Institute Chief Scientist Mike MacCracken; John McNeill of Georgetown University; former CIA Director James Woolsey; former Chief of Staff to the President John Podesta; former National Security Adviser to the Vice President Leon Fuerth; Jessica Bailey, Sustainable Development Program officer at the Rockefeller Brothers Fund; Rand Beers, president of the National Security Network; General Counsel Sherri Goodman of the Center for Naval Analysis; CNAS Senior Fellow Derek Chollet; President of the Pew Center on Global Climate Change Eileen Claussen; Gayle Smith, senior fellow at the Center for American Progress; Daniel Poneman, principal of the Scowcroft Group; Senior Fellow Susan Rice of the Brookings Institution; and principal of the Albright Group Wendy Sherman.

In particular, I want to single out the senior scientist from the Pew Center on Climate Change, Jay Gulledge, whose advice, good judgment, and expertise were essential to this project. In addition, we are deeply indebted to the National Intelligence Council and the Rockefeller Brothers Fund for their generous support of the project. I am especially grateful for the assistance of Alexander Lennon and Julianne Smith of the Center for Strategic and International Studies who helped coordinate the discussions that produced an initial report on this topic, published jointly in 2007 by Center for a New American Security and Center for Stategic and International Studies. My thanks go to Bob Faherty of Brookings Institution Press who gave us the support to expand our initial publication into the much longer, more detailed version of the study that follows.

Finally, I need to thank Senior Fellow Sharon Burke of CNAS who provided a great deal of help in reviewing drafts of our chapters from the beginning to the end and all-purpose support in bringing this project to fruition; Christine Parthemore of CNAS who devoted countless hours in editing and research assistance to ensure that this publication was of the highest standards; and Whitney Parker of CNAS, who provided editing assistance as well as guidance throughout the publication process and kept all tracks running smoothly.

CLIMATIC CATACLYSM

one

National Security and Climate Change in Perspective

KURT M. CAMPBELL AND CHRISTINE PARTHEMORE

In early 2007 the group responsible for setting the "Doomsday Clock," a depiction of the risks of imminent worldwide catastrophe, cited the threat of climate change as one reason for moving its minute hand two minutes closer to midnight.[1] Although the nuclear-era clock is perhaps an imperfect depiction of the nature of the challenge posed by climate change—the cumulative impact of human activities that affect the environment versus the kind of events that lead to a sudden conflict—climate change can provide profound and urgent threats to the well-being of mankind. Yet the risk that such catastrophe may lie at this intersection of climate change and national security is not as well understood as it should be, despite decades of exploration of the relationship between the two fields. The overall purpose of this book is to fill this gap: to provide a primer on how climate change can serve to undermine the security of the planet.

For most of 2006 and 2007, a diverse group of experts, under the direction and leadership of the Center for a New American Security (CNAS) and the Center for Strategic and International Studies (CSIS), met regularly to start a new and important conversation about this security-and-climate-change nexus and to consider the potential future foreign policy and national security implications. Our collaboration engaged climate scientists and national security specialists in a lengthy dialogue on the security implications of future climate change. As one notable scholar intoned more than a decade ago, it is necessary for such diverse professionals to "acquire detailed knowledge of a daunting range of disciplines, from atmospheric science and agricultural hydrology to energy economics and international relations theory."[2] His advice has largely been ignored, and even our eclectic group occasionally struggled to "speak the same language." But a shared sense of purpose helped us develop a common vocabulary and mutual respect, and begin the daunting process of closing these knowledge gaps among us.

A distinguished group of nationally recognized leaders was identified and recruited from the fields of climate science, foreign policy, political science, oceanography, history, and national security to take part in this endeavor. Members of the group included Thomas Schelling, the Nobel laureate in economics in 2005; Jay Gulledge, senior scientist, Pew Center on Global Climate Change; Ralph Cicerone, president of the National Academy of Sciences; Bob Correll, fellow of the American Meteorological Society; Terrence Joyce, senior scientist, and Richard Pittenger, former vice president, Woods Hole Oceanographic Institution; Mike MacCracken, chief scientist, Climate Institute; John McNeill, professor of history, Georgetown University; James Woolsey, a former director of the CIA; John Podesta, chief of staff of President Bill Clinton; Leon Fuerth, national security adviser to Vice President Al Gore; Jessica Bailey, sustainable development program officer, Rockefeller Brothers Fund; Rand Beers, president, Valley Forge Initiative; Sherri Goodman, general counsel, Center for Naval Analysis; Derek Chollet, senior fellow, Center for a New American Security; Eileen Claussen, president, Pew Center on Global Climate Change; Gayle Smith, senior fellow, Center for American Progress; Daniel Poneman, principal, the Scowcroft Group; Susan Rice, senior fellow, the Brookings Institution; and Wendy Sherman, principal, the Albright Group.

The mandate of the exercise was, on its face, very straightforward: employ the best available evidence and climate models, and imagine three future worlds that fall within the range of scientific plausibility. Such scenario planning is more than a creative writing exercise: it is a tool used successfully by businesses and governments all over the world to anticipate future events and plan more wisely in the present. The scenarios in this report use the time frame of a national security planner: thirty years, the time it takes to get major military platforms from the drawing board to the battlefield. The exception is the third, catastrophic, scenario, which extends out to a century from now.

Although the intersection of climate change and national security has yet to be fully mapped, there is a long, rich history of scholars and strategists exploring this territory. We felt it was important to begin this volume by examining this literature, in order to understand how we might begin to build on and depart from the existing intellectual framework and why the challenge of climate change remains unresolved.

Beyond the Cold War: Redefining Security

Although traditionally considered to be primarily a domestic policy concern, discussion of the environment and climate change as national security and foreign policy matters trickled through the 1970s and early 1980s. George Kennan wrote in *Foreign Affairs* in 1970 of the global scale of such issues and suggested the need for an independent international institution to track and coordinate information on what nations, states, and communities did to impact the environment.[3] In 1974 General Maxwell Taylor suggested creating "an expanded National Security Council charged with dealing with all forms of security threats, military and nonmilitary, and having access to all elements of government and to all relevant resources capable of contributing to this broad task." Taylor criticized the NSC for generally ignoring the environment and many other issues.[4] The environmentalist Lester Brown of the World-watch Institute wrote in a seminal 1977 paper, "Redefining National Security," that "threats to security may now arise less from the relationship of nation to nation and more from the relationship of man to nature. Dwindling reserves of oil and deterioration of the Earth's biological systems now threaten the security of nations everywhere."[5] In the late 1980s Egypt's Foreign Minister Boutros Boutros-Ghali warned that the next war in the Middle East would be over water.

Although the concept of conflict over natural resources has long been a strong theme in the public imagination, especially concerning water and oil, conflict related to climate change has long remained a relatively obscure topic. This changed as the threat of the cold war waned, and as carbon loading from a host of developed and developing states increased dramatically in the late 1980s and into the new century.

The Canadian government held the first major international conference focused on climate change, "The Changing Atmosphere: Implications for Global Security," in Toronto in the early summer of 1988. At that conference, Norway's prime minister, Gro Harlem Brundtland, declared, "We are now realizing that we may be on the threshold of changes to our climate, changes which are so extensive and immediate that they will profoundly affect the life of the human race." Scientists offered projections of possible temperature and sea level increases, and politicians from more than forty countries outlined security, economic, and political consequences of such changes in nature. Representatives of the host nation's government recommended that NATO and other economic and military organizations should be studied as models for international cooperation to combat climate change. However, many participants retained a policy focus of only voluntary solutions.[6]

That year, the United Nations and the World Meteorological Organization established the Intergovernmental Panel on Climate Change; its purpose was to be an independent entity to collect and analyze climate information from around the world, identify weaknesses and gaps in climate and environmental knowledge, and identify what scientific evidence government leaders required to make sound policy. The international community recognized the need for undeniable science rooted in global observations if decisionmakers were to take the threat of global warming seriously and initiate appropriate action.

In the summer of 1988, near-record temperatures and severe drought helped to spark political and popular interest in the United States as well as Canada. James Hansen, the director of NASA's Goddard Institute of Space Studies, testified before the Senate that there was a 99 percent certainty that the climate was indeed changing as a result of human contributions of greenhouse gases to the atmosphere. "It is time to stop waffling so much and say that the evidence is pretty strong," he declared, warning, "Global warming . . . is already happening now."[7]

Hansen's testimony is regarded as a major catalyst for getting Washington to think about climate change. However, it also triggered vociferous reaction from global warming skeptics of all stripes, from those who simply thought there was not yet enough data on the dynamics of clouds or the interactions between atmosphere and oceans to draw firm conclusions, to those who objected to the very concept that human activity could affect global climate patterns. For example, the climatologist Patrick Michaels responded to the uptick in warnings about climate change in a January 1989 *Washington Post* op-ed. "Of the hundred-odd scientists in the world actively involved in the study of long-term climate data, only one—James Hansen of NASA—has stated publicly that there is a 'high degree of cause and effect' between current temperatures and human alteration of the atmosphere," he wrote.[8] Hansen was forced to defend himself, responding in the *Post* the following month, "The evidence for an increasing greenhouse effect is now sufficiently strong that it would have been irresponsible if I had not attempted to alert political leaders."[9]

The Worldwatch Institute, an environmental policy research center, was a leading force in pushing the dialogue about the potential global implications of climate change. Its 1988 *State of the World* report stated: "For four decades, security has been defined largely in ideological terms. . . . The threat posed by continuing environmental deterioration is no longer a hypothetical one." As one author noted, "Threats to human security are now seen much more in environmental and economic terms and less in political ones."[10] Michael Oppenheimer, a prominent atmospheric scientist with the Environmental Defense Fund, summarized the attitude of many: "Any race of animals able to predict the warming of the Earth 100 years ago should be clever enough to stop it."[11] But the question of what to do continued to loom large, and many of the answers that were offered found no strong backing in Washington.

Of course, when it rains, it pours, and a flood of debate came about in 1989 and continued into the early nineties. As scientific evidence on climate change grew and the Soviet Union fell, an opening was created for redefinition of a new national security paradigm. The notion of elevating climate change and the environment to the level of a national security threat spread into the wider foreign policy community, instigating a heated debate.

"The 1990s will demand a redefinition of what constitutes national security," wrote Jessica Tuchman Mathews, then vice president of the World Resources Institute, in the spring 1989 issue of *Foreign Affairs,* in an article titled "Redefining Security," still credited with sparking this debate in earnest. "In the 1970s the concept was expanded to include international economics," she wrote. "Global developments now suggest the need for another analogous, broadening definition of national security to include resource, environmental and demographic issues." She described the key issue: "Environmental strains that transcend national borders are already beginning to break down the sacred boundaries of national sovereignty." She also lamented the inability of current international relationships to manage the coming environmental and climate problems: "No one nation or even group of nations can meet these challenges, and no nation can protect itself from the actions—or inaction—of others. No existing institution matches these criteria."[12]

That spring, Senator Al Gore, one of the more vocal politicians adding weight to the climate change debate, expanded on this notion. "As a nation and a government, we must see that America's future is inextricably tied to the fate of the globe," he wrote. "In effect, the environment is becoming a matter of national security—an issue that directly and imminently menaces the interests of the state or the welfare of the people."[13]

Journalists, politicians, editorial boards, and scientists soon began to echo the concept: climate and environmental issues are of highly relevant national security and foreign policy concern. In the summer of 1989, the G-7 summit in Paris even focused on the environment, the first time the issue was a central discussion point for the group. Though it marked a positive trend that the topic was brought to the table, President George H. W. Bush and his counterparts were criticized for not pledging strong, immediate action.

Some members of the first Bush administration were lambasted for their skepticism regarding the strength of scientific evidence of climate change, and for their adamancy in sticking to that premise in international meetings, but others in government at the time advocated the elevation of this and other environmental issues. Thomas Pickering, Bush's ambassador to the United Nations, warned that "ecoconflicts" could become a major problem in North-South tensions.[14] Bush chose a former World Wildlife Fund director, William Reilly, to head the Environmental Protection Agency (EPA), and halted the strangling of its budget that had occurred under Ronald Reagan. Reilly asserted that "ecological integrity is central to any definition of national security"[15] and proposed an array of measures to combat climate change, including improving vehicle fuel efficiency, increasing solar power research, and creating fees to deter coal and oil use, but to little avail. In the same month, before James Hansen was to testify again to the Senate on global warming, executive branch officials altered his testimony to cast doubt on his own scientific judgment. Congress and the press learned of this before the hearing, sparking a huge backlash at the skepticism and stalling of many members of the Bush administration concerning climate change.[16]

This trend—two contradictory responses to climate change—continued as some in the U.S. government began to treat environmental issues and climate change as strategically important fields, while others, including many in senior positions, pushed back or outright rejected the notion. The debate created enough waves to warrant significant attention from the mainstream press. In October 1989 *Time* magazine indicated which side of the debate seemed to be taking the lead in a special report, "The Greening of Geopolitics," with the headline, "A New Item on the Agenda: The Plight of the Planet Is Finally Serious International Business."[17]

The debate carried over into military considerations as well. Technologies that had been designed for military use or by the military were used to detect climate patterns, and old intelligence was opened for use in evaluating atmospheric data. Senator Sam Nunn, chairman of the Armed Services Committee, articulated in 1990 that "a new and different threat to our national security is emerging—the destruction of our environment. The defense establishment has a clear stake in countering this growing threat. I believe that one of our key national security objectives must be to reverse the accelerating pace of environmental destruction around the globe."[18]

Secretary of State James Baker's FY 1991 budget request testimony restated plainly that nontraditional threats, including environmental ones, were of national security concern:

> Today and in the future, we must take collective responsibility for ensuring the safety of the international community. Traditional concepts of what constitutes a threat to national and global security need to be updated and extended to such divergent concerns as environmental degradation, narcotics trafficking, and terrorism. Our nonrenewable resources, human lives, and the values of civilized society all are irreplaceable assets which we cannot fail to protect.[19]

But climate change specifically was still treated by most as a very separate track from environmental concerns more broadly. Many who pushed the misconception that the science behind climate change was preliminary and that evidence of the link between emissions of carbon dioxide (CO_2) and global warming was inconclusive only pushed harder into the nineties. EPA administrator William Reilly continued to warn of the dangers of a failure to act, stating in 1992: "We invested so much in responding to [a possible] nuclear attack from the USSR, even though the risk may not have been that high. . . . The risk of climate change is so much larger and yet there has been no equivalent thinking to insure ourselves against it."[20]

The Clinton Years

Global warming and environmental issues in general became one of several major policy areas of focus during Bill Clinton's presidency. Vice President Al Gore and the first undersecretary of state for global affairs, Timothy Wirth, who as Senate colleagues had been two of the leading advocates of action on climate change, signified a wave of change within the government. The cold war was over, and environmentalists advocated using an expected "peace dividend" to halt climate change and ozone depletion. The threat of climate change was often juxtaposed with the nuclear threat, with considerable argument as to which posed the greater danger.

As recognition spread that national security needed to be redefined to encompass threats not strictly military, the focus shifted primarily to economic and demographic issues. Perhaps this was because it was more direct and obvious *why* and *how* these areas generated significant threats, and because it was more readily apparent how the nation could handle these challenges. This trend snowballed as not only climate change skeptics but also some who agreed that global warming was a challenge of high importance began to portray economic growth and environmental regulation as antithetical. Warnings of a recession and a perception that countering global warming entailed extreme expenses further stalled its moving to a place of high national priority.

In late 1993 Clinton unveiled a Climate Change Action Plan, a series of voluntary measures to reduce greenhouse gas emissions to 1990 levels by 2000. The plan both disappointed many environmentalists for not creating a system of mandatory measures and was praised by others for taking the health of the economy into account. Energy Secretary Hazel O'Leary declared that inaction would be met with the enactment of mandatory measures, which encouraged many in the middle to consider the plan a good first step in slowing the pace of global warming.[21] A few months later, Undersecretary Wirth addressed the link between climate change and national security at a United States Information Agency foreign press briefing: "We're working on—continue to work on—global climate change. The U.S. has put together its action plan." At the same time he signaled that the Clinton administration did not consider global warming as a single, paramount threat: "We have very broad support in the Congress for this [the administration's agenda] in the post–Cold War era as the United States redefines its examination of national security. As the president pointed out . . . this falls into three broad categories of nuclear non-proliferation, focuses on democracy, and sustainable development."[22] Climate change was merely a subheading.

In a 1994 *Atlantic Monthly* article on demographic and environmental issues creating anarchic conditions in Africa that is still cited to this day as another major catalyst for attention to the topic, Robert Kaplan implored, "It is time to understand 'the environment' for what it is: *the* national-security issue of the early twenty-first century" (emphasis in original).[23] But policymakers never elevated it to this level. Press focus also changed, as climate change waned as a political hot topic and morality, globalization, and technology took increasing command of the national conversation. The environment and security scholar Geoffrey Dabelko observed in 1999 that the "bubble burst" after 1994, and "the policy crowd moved on to other theories about the roots of conflict. Ethnicity and 'the clash of civilizations' . . . now claimed the spotlight."[24]

But although academics and policy wonks may have dropped the serious debate over whether the environment and climate change were national security concerns per se, climate change nevertheless stood in the foreign policy spotlight in Clinton's second term. The 1997 Kyoto Protocol, the international agreement that required the reduction of greenhouse gases to below 1990 levels between 2008 and 2012 on the part of the developed nations that ratified it, became a source of tension throughout much of Clinton's second term.

Greenhouse gas emissions increased through the 1990s, as the global economy boomed with new players such as China on the international scene, and many viewed Kyoto as merely a first step that might reduce the rate of increase of emissions, but not knock them back to earlier levels. For the United States, Clinton proposed a system to cap emissions and a system of trading emissions credits, along with funding research and development through tax credits. Clinton's initial proposal guaranteed no serious action for nearly a decade, and reports trickled out that stronger policies advocated by his environmental advisers were systematically weakened on the advice of administration economists; this meant that the United States would be going into the Kyoto negotiations from this tempered position. The president warned that he was prepared to reject stricter standards demanded by European and other nations and threatened that he would not submit the treaty for Senate approval, with the stated reasoning that developing nations would not be required to comply.[25]

The United States did sign the Kyoto Protocol in 1998, but the *how* of an international system of reducing emissions was a point of great contention. The United States wished to count increased and protected forest and agricultural land as carbon sinks. Other developed nations took this as an easy way out, and this—along with discord over compliance monitoring, enforcement, and the question of which nations would bear what costs—led to the collapse of the Kyoto negotiations by late 2000, the year before ratification was to occur.[26] Clinton never did send the Kyoto treaty to the Senate for approval, but the years of debate over multiple sticking points made it clear that to do so would have been fruitless anyway. The question of whether the United States would agree to international environmental standards was effectively answered by both Democrats and Republicans with a resounding "no."

The New Millennium

Just two months into George W. Bush's presidency, his EPA administrator announced that the administration had no intention of implementing the Kyoto treaty. Bush's reservations echoed Clinton's—it might stall economic growth, and developing nations such as China and India were not required to comply—and cast doubt on the scientific evidence that human activity drove climate change.[27] In March 2001 Bush also wrote in a letter to Republican senators, "We must be very careful not to take actions that could harm consumers. . . . This is especially true, given the incomplete state of scientific knowledge of the causes of, and solutions to, global climate change and the lack of commercially available technologies for removing and storing carbon dioxide."[28] One hundred and seventy-five countries eventually accepted the treaty, and most developed nations ratified it, but not the United States.

Years of government inaction on climate change followed. Bush enjoyed years with a Republican Congress under his strict instruction. The September 11 terrorist attacks, Bush's ambiguous and ambitious Global War on Terror, and two wars distracted attention and funds from virtually all else. Perhaps because the concept of a threat was now painted in such stark terms—attack on the American mainland, and anything that might enable it—a debate over the environmental links to national security was sparked anew, if under the mainstream radar. And though the problem was once compared to the nuclear threat, its comparison with and linkage to terrorism now became prominent. In a 2005 article titled "Climate Change Poses Greater Security Threat than Terrorism," Janet Sawin of Worldwatch Institute asserted that transformations in the climate would disrupt global water supplies and agricultural activities, resulting in drought and famine, which would lead some people to turn to extralegal organizations and terrorist groups that would be able to provide for their basic needs better than existing economic and political institutions.[29]

The momentum to discuss the impact of climate change as a national security issue has finally been building steadily since 2006. Al Gore's climate change slide show, *An Inconvenient Truth,* became both a best-selling book and a documentary film that won multiple awards, including an Oscar. Gore and the Intergovernmental Panel on Climate Change were awarded the 2007 Nobel Peace Prize. Television news programs and presidential candidates hosted "green weeks" to showcase information about climate change and ideas on how to address it. Some in the mainstream press admitted that they long lent narrow special interests too much credence, and thousands of scientists around the world too little. The prominent *New York Times* columnist Nicholas Kristof has publicly spoken out about what he sees as a failure of the elite media to cover the issue of climate change in all of its manifestations. Climate change rather than the perennial issues of globalization, nuclear proliferation, and the Iraq War dominated the January 2007 World Economic Forum meeting of the world's political and business leaders in Davos, Switzerland.[30] In explaining why he chose to discuss climate change at Davos, the leader of Britain's Conservative Party, David Cameron, said, "There is a consensus . . . that says we need to take action to prevent it, rather than just mitigate its effects. But, at the same time, politicians have a duty to prepare for its consequences in terms of domestic and international security."[31] Policy leaders and academics are now airing and debating concepts for post-Kyoto international cooperation.

The Long-standing Debate Concerning Redefinition

Remarkably, the debate as to whether or not it is appropriate to define climate change in particular and the environment in general as national security concerns still continues today. A primary source of disagreement between the camps as to whether national security should be redefined is the question of whether a security framing implies military, rather than political, solutions. Examples of reasoning from both sides of the debate show how nuanced the intellectual wrangling has become.

Daniel Deudney of Princeton University wrote in 1991 that "such experimentation in the language used to understand and act upon environmental problems is a natural and encouraging development. But not all neologisms and linkages are equally plausible or useful." He argued that environmental and security threats were so inherently different in nature that they should not be linked and contended that threats of all kinds were likened to war to arouse emotional responses and create a sense of urgency, regardless of the appropriateness of such connections.[32]

The environmental scholar Peter Gleick, in the same publication, took a similar tack of rejecting a broadened definition of national security, but with different justification. Citing the Persian Gulf War as an example, Gleick posited that in the future, threats diversely political, military, and environmental in nature would become "more tightly woven" and inherently all pieces of a single greater picture. He suggested that "what is required is not a redefinition of international or national security, as some have called for, but a better understanding of the nature of certain threats to security, specifically the links between environmental and resource problems and international behavior." He emphasized that threats to the environment will inevitably have to be of concern in the future, and "climatic changes are most likely to affect international politics."[33]

However, as the scholar Joseph Romm wrote of the time after the collapse of the Soviet empire, the reason for broadening the definition of national security is to reorient leaders and the established bureaucracy toward seeking and considering a broader array of solutions: "Our existing security paradigm is increasingly inadequate to address these problems."[34] In 2000 the *New York Times* reporter David Sanger described the Clinton administration's expanded national security definition: "In [Al] Gore's case, the political calculation is obvious: he is trying to portray his opponent, George W. Bush, as a man caught in what he termed a cold-war mindset. . . . It is possible—though hardly certain—that after a decade of unchallenged power and prosperity, Americans are ready to think about their national security in broader terms."[35]

But while academics, analysts, and the media could argue over definition until the end of time, it is still up to elected national leaders to enact serious change. The only way to viably do so is to settle the definitional issue once and for all, by convincing the American people that climate change is perhaps the top national security problem we face.

Predicted National Security Consequences of Climate Change

The range of scholars' predictions of the consequences of climate change and severe environmental degradation has remained largely consistent over the past thirty to forty years. In addition to the diverse speculation over how serious the effects of climate change will be, such predictions cover a broad spectrum of mild to extreme human reaction to the repercussions of global warming, such as sea level rise and altered agricultural productivity.

In his pivotal 1977 paper, "Redefining National Security," Lester Brown wrote that "excessive human claims" on the environment threatened nearly all aspects of life: fishing and crop yields, forest regeneration, economic stability, and energy production and use.[36] Ian Rowlands, of the London School of Economics, intoned in *Washington Quarterly* in 1991 that "no country will be immune from the security challenges posed by global environmental change." Moreover, he described it as a unique issue in that the threat was not external: the behavior of the United States and other nations constituted a threat to themselves, and security would not be dependent upon the actions of any single player.[37] In the same publication, the environmental consultant A. J. Fairclough wrote that natural resources would become increasingly scarce in the future, aggravating existing tensions and creating new threats of economic stagnation and refugee flows. He summarized: "These threats to environmental security must be seen as threats to the well-being and quality of life of our populations that are every bit as serious as military threats. We need to react accordingly."[38] The High-Level Panel on Threats, Challenges, and Change appointed by former UN secretary general Kofi Annan warned in 2004 of a vicious cycle of poverty, disease, environmental degradation, and civil violence.[39]

Scholars and policymakers have long spotlighted migration, both within and between nations, as an early and pervasive consequence of climate change. In April 1974 the president's National Security Council compiled a National Security Study memo for President Richard Nixon, unique in its consideration of population pressures as a potential threat to national security. This study was quite prescient, as many of the population trends it described came to the fore in the Clinton years and remain of concern in today's debates on global warming. "Population factors appear to have had operative roles in some past politically disturbing legal or illegal mass migrations, border incidents, and wars," the study stated. "If current increased population pressures continue they may have greater potential for future disruption in foreign relations."[40]

One of the most noted theorists of environment-conflict studies, Thomas Homer-Dixon, wrote that change to the environment will impact populations by "decreased economic productivity and disrupted institutions will jointly contribute to relative-deprivation conflicts . . . positive feedbacks may operate: relative-deprivation conflicts may cause further economic decline and institutional dislocation."[41] Joseph Romm echoed this concern in 1993, writing that many nations' being confronted with scarce resources "may lead to conflict or ecosystem collapse, resulting in environmental refugees. Such traumas could threaten U.S. national security if these conflicts were to occur in areas of importance to the United States, or if refugees were to flee in large numbers to this country."[42] And the German climatologist Hermann Ott wrote in 2001, "Water and food shortages, rising sea levels and generally changing patterns of precipitation will lead to mass migrations and a considerable increase in low- and high-intensity warfare in many parts of the southern world."[43]

A group of scholars who used statistical and quantitative methods to track population growth, agricultural production, global climate changes, and war found the heaviest correlations among these factors in arid regions. Basing their assessments on the overlap of changing climate patterns and conflict, they wrote that "the greater threat from global warming comes from uncertainty of the ecosystem change," for that uncertainty will cause social and economic turmoil and other secondary effects wherever quick adjustment cannot be made. "A change of one key component under global warming would likely cause disastrous results in human societies dependent on the existing human ecosystem." These scholars posed key questions: "Is the changed ecosystem sufficiently adaptable or are the adaptation choices affordable for all of us?"[44]

One-third of the world's population lives within 60 kilometers (about 37 miles) of a coastline, so potential refugee crises are of critical concern if the widespread sea level rises predicted by scientific models of global warming occur. The sheer numbers of potentially displaced people are staggering. One recent World Bank report included calculations that over the course of the twenty-first century, sea level rise due to climate change could displace hundreds of millions of people in developing countries.[45] Christian Aid and other nongovernmental organizations have estimated that climate change could deprive as many as 1 billion people of their homes between now and 2050.[46] A two-day conference in Oslo in the summer of 2005, the International Workshop on Human Security and Climate Change, was dedicated to evaluating how climate change will drive human migration, as scholars around the world are struggling to determine how to best cope with such trends.

Among the long litany of devastating predicted effects, one focus of increasing concern is the disproportionate harm to the world's poorest people. "In low-income countries unable to offset crop shortfalls with imports, a production drop can translate directly into a rise in death rates," Lester Brown wrote in 1977.[47] Norway's Prime Minister Gro Harlem Brundtland warned at the 1988 Toronto climate change conference, "Climatic change will affect us all profoundly, regardless of where we live. And, as always, the poorest countries will be the ones most severely affected."[48] Jessica Tuchman Mathews also made the point that although some areas might benefit from better agricultural conditions, all regions will be susceptible to highly variable and unpredictable changes. Further, adapting to climate change "will be extremely expensive. Developing countries with their small reserves of capital, shortages of scientists and engineers, and weak central governments will be the least able to adapt, and the gap between the developed and developing worlds will almost certainly widen."[49]

Many of the predicted consequences of global warming and environmental change are already occurring. Thomas Homer-Dixon's 1991 list of expected consequences stands out:

> Environmental change may contribute to conflicts as diverse as war, terrorism, or diplomatic and trade disputes. Furthermore, it may have different causal roles: in some cases, it may be a proximate and powerful cause; in others, it may only be a minor and distant player in a tangled story that involves many political, economic, and physical factors. . . . Warmer temperatures could lead to contention over new ice-free sea-lanes in the Arctic or more accessible resources in the Antarctic.[50]

Sixteen years later, in August 2007, a Russian adventurer descended 4,300 meters (about 15,000 feet) under the thinning ice of the North Pole to plant a titanium flag and claim some 1.2 million square kilometers (468,000 square miles) of the Arctic for Russia. Not to be outdone, the prime minister of Canada stated his intention to boost his nation's military presence in the Arctic, with the stakes raised by the recent discovery that the heretofore iced-in Northwest Passage has become navigable for the first time in recorded history. Elsewhere on the globe, the spreading desertification in the Darfur region has compounded the tensions between nomadic herders and agrarian farmers, providing the environmental backdrop for genocide.

What Can Be Done?

Whatever the possible international distribution of climate change effects, there is a general consensus about the need for multilateral cooperation, for isolated and uncoordinated national-level steps are clearly not up to the task. In the October 2006 *Review on the Economics of Climate Change,* the former World Bank economist Nicholas Stern maintained that although the near-term costs of stabilizing the concentration of greenhouse gases in the atmosphere are significant but manageable (approximately 1 percent of global GDP), any major delay in responding would result in substantially higher aggregate costs, amounting to an estimated loss of up to 20 percent of the world's GDP. One of the report's key assessments is that all countries can contribute to combating climate change while still achieving economic growth. In particular, the Stern review urged a multidimensional international response involving expanded use of carbon emissions trading arrangements, increased cooperation in developing and sharing low-carbon technologies, curbing deforestation, and greater support for adaptation measures.[51]

At the end of March 2007, the U.S. Army War College sponsored a two-day conference at the Triangle Institute for Security Studies on the topic "The National Security Implications of Global Climate Change." Participants included civilian strategists and active-duty and retired military officers, who explored a range of issues potentially linking climate change to international security. A major goal of the conference was to assess how the military could mitigate climate change, assist in efforts to adapt to climate change, and prepare for the security challenges that might ensue from climate change. The attendees stressed that any effective response to climate change–related security problems likely would require multi-agency cooperation, especially for domestic emergency management, and typically multinational action.[52]

In April 2007 the Center for Naval Analysis (CNA) Corporation issued a landmark report that attracted major attention in the national security community because of its advisory board of former senior U.S. military officers.[53] The authors recognized that much scientific uncertainty regarding climate change persists but urged "moving beyond the argument of cause and effect," since observed climate change was already occurring and presenting challenges to national security planners. According to the report, "The chaos that results can be an incubator of civil strife, genocide, and the growth of terrorism." The authors warn that these developments could contribute to state failure, interstate conflicts, or other security problems in many geographic regions that could require a response by an already overburdened U.S. military. Transformations in the environment resulting from climate change could also complicate regular U.S. military operations. Hurricanes and rising sea levels could threaten U.S. military facilities, extremely hot or cold weather could disrupt U.S. military operations, and allied militaries might offer less support for joint missions if they also have to respond to environmental threats. The board affirmed that they as military officers had long recognized the need to assess the risks of climate change–driven events if the consequences could prove sufficiently severe.

In the face of these challenges, the CNA panel recommended that the United States adjust its national security and national defense strategies to account for the possible consequences of climate change.[54] For example, the Department of Defense should conduct an impact assessment of how rising sea levels, extreme weather events, and other effects of climate change might affect U.S. military installations over the next three to four decades. They also cautioned that extreme environmental conditions degrade weapons systems and the capabilities of military personnel. Beyond the military dimension, the panel members urged that the U.S. government seek to enhance the resilience of the international community in the face of climate-related threats by strengthening the governance, health care, and disaster prevention and relief capabilities of foreign countries. They noted that the recent creation of the U.S. Africa Command (AFRICOM) seems to serve such a purpose. The authors also recommended that the United States help limit climate change through unilateral and multilateral measures; the Department of Defense could contribute through more efficient use of energy and other measures.

STEPPING FROM THE foundation these previous works have laid, the working group convened by the Center for a New American Security and the Center for Strategic and International Studies diverged from previous work by looking into the past for historical evidence of what may be to come, then employing the best available evidence and climate models to ponder the national security implications of three plausible future worlds.

These particular scenarios aim not to speculate centuries into the future, as some scientific models do, but to consider possible developments using a reasonable time frame for making acquisition decisions or judgments about larger geopolitical trends. In national security planning, it generally can take about thirty years to design a weapons system and bring it to the battlefield, so it is important to anticipate future threat environments and prepare for the challenges we may face as a result of climate change.

The three scenarios we develop in this study are based on *expected, severe,* and *catastrophic* climate cases. The first scenario projects the effects in the next thirty years with the *expected* level of climate change. The second, *severe,* scenario posits that the climate responds much more strongly to continued carbon loading over the next few decades than predicted by current scientific models. It foresees profound and potentially destabilizing global effects over the course of the next generation or more. Finally, the third, *catastrophic,* scenario is characterized by a devastating "tipping point" in the climate system, perhaps fifty to one hundred years hence. In this future world, radical changes in global climate conditions include the rapid loss of the land-based polar ice sheets, an associated dramatic rise in global sea levels, and the destruction beyond repair of the existing natural environment.

For each of the three plausible climate scenarios, we asked a national security expert to consider the projected environmental effects of global warming and to map out the possible consequences for peace and stability. We also enlisted a historian of science to consider whether there was anything to learn from the experience of earlier civilizations confronted with rampant disease, flooding, or other forms of natural disaster. Each climate scenario was carefully constructed and the three corresponding national security futures were thoroughly debated and discussed by the group. A synthesis and summary of some of the key findings from the various chapters follows.

Historical comparisons from previous civilizations and national experiences of such natural phenomena as floods, earthquakes, and disease may be of help in understanding how societies will deal with unchecked climate change. In the past, natural disasters generally have been local or abrupt or both, making it difficult to directly compare the worldwide effects of prolonged climate change to historical case studies. No precedent exists for a disaster of this magnitude—one that affects entire civilizations in multiple ways simultaneously. Nonetheless, the historical record can be instructive; human beings have reacted to crisis in fairly consistent ways. Natural disasters have tended to be divisive and sometimes unifying, to provoke social and even international conflict, to inflame religious turbulence, focus anger against migrants or minorities, and direct wrath toward governments for their actions or inaction. People have reacted with strategies of resistance and resilience—from flood control to simply moving away. Droughts and epidemic disease have generally exacted the heaviest toll, in both demographic and economic terms, and both are expected effects of future climate change. Indeed, even though global warming is unprecedented, many of its effects will be experienced as local and regional phenomena, suggesting that past human behavior may well be predictive of the future. This history is explored in chapter 2.

In chapter 3, the climate scientist Jay Gulledge explains how projections about the effects of climate change have tended to focus on the most *probable* outcome, on the basis of mathematical modeling of what we know about the global climate. With climate science, however, the level of uncertainty has always been very high. Indeed, the scientific community has been shocked at how fast some effects of global warming are unfolding, which suggests that many of the estimates considered most probable have been too conservative.[55] When we build climate scenarios in order to anticipate the future, therefore, there is a very strong case for looking at the full range of what is *plausible*, and this task is taken for the effects of climate change by region.

The *expected* climate change scenario considered in this report, an average global temperature increase of 1.3°C (2.3°F) by 2040, can be reasonably taken as a basis for national planning. The authors of chapter 4 write that the environmental effects in this scenario are "the least we ought to prepare for." National security implications include heightened internal and cross-border tensions caused by large-scale migrations; conflict sparked by resource scarcity, particularly in the weak and failing states of Africa; increased disease proliferation, which will have economic consequences; and some geopolitical reordering as nations adjust to shifts in resources and prevalence of disease. Across the board, the ways in which societies react to climate change will refract through underlying social, political, and economic factors.

In the case of *severe* climate change, corresponding to an average increase in global temperature of 2.6°C (4.7°F) by 2040, massive nonlinear events in the global environment give rise to massive nonlinear societal events. In this scenario, discussed in chapter 5, nations around the world will be overwhelmed by the scale of change and the accompanying huge challenges, such as pandemic disease. The internal cohesion of nations, including the United States, will be under great stress, as a result both of a dramatic rise in migration and of changes in agricultural patterns and water availability. The flooding of coastal communities around the world, especially in the Netherlands, the United States, South Asia, and China, has the potential to challenge regional and even national identities. Armed conflict between nations over water resources such as the Nile and its tributaries is likely, and nuclear war is possible. The social consequences range from increased religious fervor to outright chaos. In this scenario, climate change provokes a permanent shift in the relationship of humankind to nature.

The *catastrophic* scenario, with average global temperatures increasing by 5.6°C (10.8°F) by 2100, is by far the most difficult future to visualize without straining credulity. The author of this scenario notes that intense hurricanes will become increasingly common, and so will droughts, floods, wildfires, heat waves, and churning seas. Hundreds of millions of thirsty and starving people will have to flee these disasters or perish, leaving the globe dotted with ghost towns. The abrupt and sudden nature of many of these phenomena will challenge the ability of all societies to adapt. If the catastrophic scenario described in this chapter comes to pass, the world will be caught in an age where sheer survival is the only goal.

The author of chapter 7 describes climate change as a malignant rather than a malevolent threat. However, one of the other great challenges of our time, terrorism, a malevolent threat, provides surprisingly similar challenges to national security as does climate change. Both threaten the economy and the ways the United States uses energy, and both contribute to the vulnerability of the nation's critical infrastructure. In this sense, diverse groups whose interests center on either the environment or on national security have cause to come together and act in tandem.

The Kyoto Protocol will expire in 2012. The United States, the European Union, and China are responsible for roughly half of global greenhouse gas emissions. The authors of chapter 8 outline changing views in each of these powerhouse nations, noting that if just these three players can agree to reduce greenhouse gas emissions, and especially if experts from all relevant communities participate, a post-Kyoto climate change framework is likely.

The United States must confront the harsh reality that unchecked climate change will come to represent perhaps the single greatest risk to our national security, even greater than terrorism, rogue states, the rise of China, or the proliferation of weapons of mass destruction. The effects of climate change will further complicate most other security threats as well.

Though the environment-security link has long been debated in different policy circles, scientists and national security practitioners have only recently begun to work together. New dialogues among these communities and scenario planning based on the best current scientific data have begun to paint a harrowing and Hobbesian view of the earth's future. A climatically disrupted future would be marked by sharp increases in global sea levels, endemic drought, more frequent and more extreme weather events, spreading disease vectors, massive extinctions, the prospects for the collapse of agricultural sectors and global fisheries, and the largest human movements and disruptions in history. These outcomes, on a lesser scale, will be seen even if climate change is relatively mild.

Conclusion

When one reviews years of writing, scholarly work, and government policy on climate change and national security, it is striking how well trod the ground is. This report reflects back to concepts that have been part of the national dialogue for decades. Yet even more striking is the lack of resolution. Why has serious action not been taken? After decades of warnings and years of events trending just as those warnings predicted, why has climate change not taken its rightful place as perhaps the biggest security challenge the United States faces?

Part of the explanation is political, part of it lies with the earlier uncertainty of the science, and part of it is institutional. Government bureaucracy is hard to move, as are political will and public opinion. Second, the trend continues today of labeling climate-neutral technologies "too expensive" or detrimental to economic growth. The price of change is still calculated solely in direct terms, without regard to the nonmonetary quality of life, the second-order costs of the status quo, or the security benefits that change would bring. In a prophetic 1989 *Washington Post* article, "Our Global Eco-Blindness; Earth's Fate Is the No. 1 National Security Issue," Al Gore enunciated this problem:

> The effort to solve the global environmental crisis will be complicated not only by blind assertions that more environmental manipulation and more resource extraction are essential for economic growth. It will also be complicated by the emergence of simplistic demands that development, or technology itself, must be stopped for the problem to be solved. This is a crisis of confidence which must be addressed.[56]

Even more important, the various communities at work on this topic have never truly united as a single, undeniable force for recognizing the urgent need for action. This project was formed in an attempt to play a small role in correcting this critical deficiency. It is the hope of all who contributed to this project that this volume will provide a basis for understanding and action. There are disagreements in all areas of national security, and counter-arguments that challenge prevailing orthodoxies are not treason. But in all manner of security concerns the national leadership is accorded broad latitude to act to protect the country. Climate change must come to be systematically accepted alongside comparable threats to the very security of the nation—and the world.

This report and the interactions among participants highlighted inescapable, overriding conclusions. We hope this study will make an important contribution to the understanding of what might well be the single most significant challenge confronting the United States—indeed, human civilization. We approached the task with humility: understanding the scope and the scale of climate change is not easy. It is even harder to come up with credible ideas and options for managing and mitigating the effects of global warming. We hope that this collaborative effort offers a strong foundation for such efforts.

Notes

1. Molly Bentley, "Climate Resets 'Doomsday Clock,' " *BBC News,* January 17, 2007.

2. Thomas F. Homer-Dixon, "On the Threshold: Environmental Changes as Causes of Acute Conflict," *International Security* 16 (Autumn 1991): 84.

3. George F. Kennan, "To Prevent a World Wasteland: A Proposal," *Foreign Affairs* 48 (April 1970) (www.foreignaffairs.org/19700401faessay48301/george-f-kennan/to-prevent-a-world-wasteland-a-proposal.html).

4. Maxwell D. Taylor, "The Legitimate Claims of National Security," *Foreign Affairs* 52 (April 1974): 575–94.

5. Lester Brown, "Redefining National Security," Worldwatch paper 14 (Washington: Worldwatch Institute, October 1977), p. 6.

6. See, for example, Pat McNenly, "Borrow NATO Model for Struggle against Pollution, Official Urges," *Toronto Star,* June 29, 1988, p. A32; Philip Shabecoff, "Norway and Canada Call for Pact to Protect Atmosphere," *New York Times,* June 28, 1988, p. C4; and Ronald Kotulak, "Pace of Ozone Loss Triggers Alarm," *Chicago Tribune,* June 28, 1988, p. 8.

7. Philip Shabecoff, "Global Warming Has Begun, Expert Tells Senate," *New York Times,* June 24, 1988, p. 1.

8. Patrick Michaels, "The Greenhouse Climate of Fear," *Washington Post,* January 8, 1989, p. C3.

9. James Hansen, "I'm Not Being an Alarmist about the Greenhouse Effect," *Washington Post,* February 11, 1989, p. A23.

10. Lester R. Brown and others, *State of the World 1988* (Washington: Worldwatch Institute, 1988), pp. xv, 182.

11. Michael Weisskopf, " 'Greenhouse Effect' Fueling Policy Makers; Concept from the 19th Century 'Is Here,'" *Washington Post,* August 15, 1988, p. A1.

12. Jessica Tuchman Mathews, "Redefining Security," *Foreign Affairs* 68 (Spring 1989): 162–77.

13. Al Gore, "Our Global Eco-Blindness; Earth's Fate Is the No. 1 National Security Issue," *Washington Post,* May 14, 1989, p. C1.

14. Glenn Garelik, "Greening of Geopolitics—A New Item on the Agenda: The Plight of the Planet Is Finally Serious International Business," *Time,* October 23, 1989, p. 60.

15. Michael Satchell, "The Whole Earth Agenda," *U.S. News & World Report,* December 25, 1989, p. 50.

16. See, for example, Cass Peterson, "Experts, OMB Spar on Global Warming; 'Greenhouse Effect' May Be Accelerating, Scientists Tell Hearing," *Washington Post,* May 9, 1989, p. A1; Philip Shabecoff, "White House Admits Censoring Testimony," *New York Times,* May 9, 1989, p. 1.

17. Garelik, "Greening of Geopolitics."

18. Rudy Abramson, "Nunn Sees Military Role in Aiding Environment," *Los Angeles Times,* June 29, 1990, p. 31.

19. U.S. Department of State, "Secretary Baker: Prepared statement before the Senate Foreign Relations Committee," dispatch, February 1, 1990.

20. Larry B. Stammer, "Geopolitical Effects of Global Heating Gauged," *Los Angeles Times,* February 10, 1992, p. 1.

21. See, for example, "Clinton Asks Help on Pollution Goal," Associated Press, October 20, 1993; Melissa Healy, "Clinton Unveils First Phase of Fast Action Global Warming Plan," *Los Angeles Times,* October 20, 1993; and "A Prudent Plan on Global Warming," *Chicago Tribune,* editorial, November 22, 1992, p. 14.

22. Timothy Wirth, "USIA Foreign Press Center Briefing: Clinton Administration's Focus on Foreign Affairs," Federal News Service, February 14, 1994.

23. Robert D. Kaplan, "The Coming Anarchy," *Atlantic Monthly,* February 1994.

24. Geoffrey Dabelko, "The Environmental Factor," *Wilson Quarterly,* September 22, 1999, p. 14.

25. See, for example, Joby Warrick, "Administration Signs Global Warming Pact," *Washington Post,* November 13, 1998, p. A26; Martha M. Hamilton, "The Price of Achieving Kyoto Goals; U.S. to Consider Range of Options, Technologies If Pact Advances," *Washington Post,* December 12, 1997, p. A41; and John H. Cushman Jr. and David E. Sanger, "Global Warming; No Simple Fight: The Forces That Shaped the Clinton Plan," *New York Times,* December 1, 1997, p. F3.

26. See, for example, Joseph Boris, "U.S., Europe Resume Talks on Global-Warming Treaty in Canada," United Press International, December 5, 2000; and Ray Moseley, "Global-Warming Meeting Fails after Last-Minute Deal Crumbles," *Chicago Tribune,* November 26, 2000, p. 6.

27. "U.S. Won't Follow Climate Treaty Provisions, Whitman Says," Associated Press, March 28, 2001.

28. Douglas Jehl and Andrew C. Revkin, "Bush, in Reversal, Won't Seek Cut in Emissions of Carbon Dioxide," *New York Times,* March 14, 2001, p. 1.

29. Janet L Sawin, "Climate Change Poses Greater Security Threat Than Terrorism," Global Security Brief 3 (Washington: Worldwatch Institute, April 2005). See also Gregory D. Foster, "A New Security Paradigm," *Worldwatch,* January–February 2005, p. 45.

30. John Zarocosta, "World Leaders Rally around Climate-Change Issue," *Washington Times,* January 29, 2007, p. A11.

31. David Cameron, "A Warmer World Is Ripe for Conflict and Danger," *Financial Times,* January 24, 2007, p. 15; Scott Baldauf, "Africans Are Already Facing Climate Change," *Christian Science Monitor,* November 6, 2006, p. 4.

32. Daniel Deudney, "Environment and Security: Muddled Thinking," *Bulletin of Atomic Scientists,* April 1991, pp. 22–28.

33. Peter H. Gleick, "Environment and Security: The Clear Connections," *Bulletin of Atomic Scientists,* April 1991, pp. 17–21.

34. Joseph J. Romm, *Defining National Security: The Nonmilitary Aspects* (New York: Council on Foreign Relations Press, 1993), p. 1.

35. David E. Sanger, "Sometimes, National Security Says It All," *New York Times,* May 7, 2000, p. 3.

36. Brown, "Redefining National Security," pp. 6–7.

37. Ian Rowlands, "The Security Challenges of Global Environmental Change," *Washington Quarterly* 14 (Winter 1991): 99.

38. A. J. Fairclough, "Global Environmental and Natural Resource Problems—Their Economic, Political, and Security Implications," *Washington Quarterly* 14 (Winter 1991): 78.

39. High-Level Panel on Threats, Challenges and Change, *A More Secure World: Our Shared Responsibility* (New York: United Nations, 2005), p. 15.

40. Stephen D. Mumford, "The Life and Death of NSSM 200: How the Destruction of Political Will Doomed a U.S. Population Policy" (memorandum text at www.population-security.org/28-APP2.html).

41. Homer-Dixon, "On The Threshold: Environmental Changes as Causes of Acute Conflict," p. 110.

42. Romm, *Defining National Security: The Nonmilitary Aspects,* p. 22.

43. Hermann E. Ott, "Climate Change: An Important Foreign Policy Issue," *International Affairs* 77, no. 2 (2001): 295.

44. David D. Zhang and others, "Global Climate Change, War, and Population Decline in Recent Human History," *Proceedings of the National Academy of Sciences* 104 (December 4, 2007): 19214–19.

45. Yan Jianping and others, "The Impact of Sea Level Rise on Developing Countries: A Comparative Analysis," World Bank Policy Research Working Paper 4136 (Washington: World Bank, February 2007).

46. Christian Aid, "Human Tide: The Real Migration Crisis," May 2007 (www.christianaid.org.uk/Images/human_tide3__tcm15-23335.pdf).

47. Brown, "Redefining National Security," p. 20.

48. Pat McNenly, "Pollutants Compared to Nuclear Threat," *Toronto Star,* June 28, 1988, p. A13.

49. Mathews, "Redefining Security," pp. 169–70.

50. Homer-Dixon, "On the Threshold: Environmental Changes as Causes of Acute Conflict," p. 77.

51. Nicholas Stern, *The Economics of Climate Change: The Stern Review* (Cambridge University Press, 2007).

52. For a description of the conference as well as videos of the presentations, see Strategic Studies Institute of the U.S. Army War College, "The National Security Implications of Global Climate Change" (Carlisle Barracks, Pa.: U.S. Army War College, March 30–31, 2007) (www.strategicstudiesinstitute.army.mil/events/details.cfm?q=82); Douglas V. Johnson II, "Global Climate Change: National Security Implications," Strategic Studies Institute Colloquium Brief (Carlisle Barracks, Pa.: U.S. Army War College, May 1, 2007). The summary memo by Johnson lists as a key insight that "while military forces have roles in disaster relief, the broader impact of serious climate change will require multinational, multi-agency cooperation on a scale heretofore unimaginable and could provide no-fault ground for global cooperation" (p. 1).

53. Center for Naval Analysis Corporation, *National Security and the Threat of Climate Change* (Alexandria, Va.: CNA, April 2007).

54. For additional recommendations along these lines, see Stephan Harrison, "Climate Change, Future Conflict and the Role of Climate Science," *RUSI Journal* 150, no. 6 (December 2005).

55. Stefan Rahmstorf and others, "Recent Climate Observations Compared to Projections," *Science* 316 (2007): 709.

56. Gore, "Our Global Eco-Blindness."

two

Can History Help Us with Global Warming?

J. R. McNeill

It is prudent, both intellectually and practically, to accept that the atmosphere and oceans are indeed warming, as the evidence tells us, and that this trend will accelerate in the decades ahead. While we do not and cannot know just how much warming will occur, nor how fast, we can safely say that the rapidity of warming, now and in all likelihood over the next decades, has few precedents in the history of the Earth and none in the history of civilization. This is true regardless of which of the three versions of the future offered in this book one prefers.

No instrumental records exist for prior episodes of climate change. The proxy evidence used for the reconstruction of climate history—measurements of fossil pollen, foraminifera, tree rings, oxygen isotopes in air bubbles trapped in ancient ice, and other tools—can give a good but not precise idea of past temperature and precipitation patterns.

Earth's climate has never been static. For the past 2.7 million years, it has shown a pattern of alternating long ice ages and shorter interglacials, governed by cycles in the Earth's orbit around the sun. The last ice age was at its height around 20,000 years ago. Its end (circa 11,000 to 6,000 years ago) was probably crucial for human history as it coincided with the emergence of agriculture in multiple locations. After that bout of warming—generally much slower than what we have witnessed in the last hundred years but not without sudden lurches now and again—global climate changed only modestly and slowly until the industrial age.[1] Although our Paleolithic ancestors did have to cope with rapid climate change from time to time, when they did so the Earth had fewer people (or hominids) than Chicago has today, and they were accustomed to migrating with their scant possessions as a matter of course. Their response to adverse climate change (as to much else) was to walk elsewhere. Since the emergence of agriculture, sedentarism, civilization, and the settlement of all habitable parts of the globe, the Paleolithic response has become more and more impractical. Thus, although there are analogues in Earth's history for the climate change now under way, there are none in human history. We have entered uncharted terrain.

Buffers, Resilience, and Nature's Shocks

As a species, we've enjoyed a run of luck in the Holocene, the geological period covering the last ten thousand years. Migration as a response to adversity has become progressively less viable, yet warming and cooling trends and attendant sea level fluctuations have remained small. Even the Little Ice Age, from approximately 1300 to 1850, amounted to an average cooling (in Europe, where the data are best) of just about 0.5°C (0.9°F). It made harvest failures more frequent in northern Europe and probably contributed to the demise of the tiny Greenland Norse settlement in the early fifteenth century. In lower latitudes, the Little Ice Age probably brought reduced rainfall and more frequent droughts—a much more disruptive experience than mild cooling or warming. But as nature's surprises go, the climate change of the Little Ice Age was modest.[2]

In the past, nature's shocks and stresses challenged all societies. In recent millennia, the most dangerous of these included epidemics, droughts, floods, earthquakes, and volcanic eruptions. Warming, cooling, and sea level changes were far down the list. Broadly speaking, these challenges came in two varieties: short, sharp shocks with durations of days, weeks, or a year or two; and long, slow stresses that played out over decades or centuries, and were often invisible to people at the time. In terms of demographic losses, epidemics were by far the most serious.[3]

Box 2-1 ranks the demographic seriousness of nature's shocks in very rough terms. The mortality figures, given only as an order of magnitude, represent the maximum, meaning 95 to 99 percent of such incidents would kill fewer people. So, for example, although there may have been a flood or even ten floods that killed more than 1 million people, this represents the worst that floods have ever done to humankind.

Approximate Maximum Mortality Levels from Nature's Shocks

Volcanic eruptions	10^4
Earthquakes	10^5
Floods	10^6
Droughts	10^7
Epidemics	10^8

The worst epidemics have killed 30 million to 100 million people, even if one considers the bubonic plague pandemic in fourteenth-century Eurasia (and possibly Africa) as a single event (a pandemic is an especially widespread epidemic, often global in scope). The most recent epidemic on such a scale, the influenza that raged from 1918 to 1919, killed perhaps 40 million (about 2 percent of the global population). The ongoing AIDS pandemic has so far killed 25 million to 30 million, about 0.5 percent of the current world population.[4] Such pandemics have been mercifully rare, but past epidemics that affected regions or single cities were not, and they routinely killed 5 to 10 percent or even more of the affected population.

Droughts, at their worst, have resulted in a few million deaths. The long history of drought is notably fuzzy, and whether or not deaths ought to be laid at drought's door is often unclear, especially for the deeper past. In the twentieth century, where the uncertainties are reduced, the deadliest droughts occurred in China from 1928 to 1931, in 1936, and in 1941, with 2 million to 5 million deaths on each occasion, generally because of starvation. The famous droughts in West Africa's Sahel region of 1967 to 1973 and again in the early 1980s each killed about 1 million people. In all probability some of the drought-induced Indian famines of the nineteenth century killed greater numbers, but the figures are in dispute.[5]

Floods, too, could kill thousands, even millions, although flood control and evacuation procedures have made a big difference in flood mortality. Since 1953 the annual average number of flood-caused deaths in India, the country most afflicted by floods, has been about 1,500. The worst flood in recent Chinese history, when the Yangtze surged in 1954, killed 30,000 people. Yangtze floods in 1931, perhaps the most costly ever, killed 1 million to 4 million, and those on the Hwang He (Yellow River) in 1887 resulted in perhaps 1 million to 2 million deaths. The great North Sea floods of December 1953 killed some 2,400 in the Netherlands, whereas earlier floods, in 1212, had killed 60,000. A 1342 megaflood in central Europe affected dozens of rivers, caused half of all the soil erosion over German lands in the past millennium, and probably drowned hundreds of thousands of people.[6] In 1927 the worst flood in U.S. history—until those caused by Hurricane Katrina in 2005 killed 243 people along the lower Mississippi River.[7]

Of the many thousands of deadly earthquakes, only ten have killed more than 100,000 people. The worst occurred in China in 1566; perhaps 800,000 died. The recent tsunami of December 2004, created by an undersea earthquake, killed 284,000, and the 2005 earthquake in Pakistan killed about 79,000. The San Francisco earthquake of 1906, the worst in U.S. history, killed about 3,000.[8]

Of the countless volcanic eruptions, only six are likely to have killed more than 10,000 people. The worst case, the explosion of Mount Tambora, on the northern coast of Sumbawa island, Indonesia, in 1815, took perhaps 92,000 lives; Krakatoa, in 1882, cost 36,000. The famous eruption of Mount Vesuvius in 79 C.E. killed about 3,600, and the worst in U.S. history, that of Mount St. Helens, Washington, in 1980, killed 57.

With the exception of the richer parts of the world since 1919, every generation everywhere lived under the threat of devastatingly lethal epidemics, floods, droughts, and other kinds of natural risks.[9]

As a result, all societies had to build resilience to nature's shocks. By and large, they did not intentionally build resilience or resistance to nature's slow-acting stresses, such as desiccation (the gradual drying of climate) or soil salinization, because these progressed too slowly to cause alarm, and often too slowly even to be noticed from one generation to the next. But resistance and resilience to the easily observable short, sharp shocks were, always and everywhere, an important priority.

Resistance and resilience are not the same thing. Resistance to flood, for example, can take the form of the construction of seawalls and dikes, as the Dutch have done for seven hundred years to keep the North Sea at bay. Resilience to flood means the capacity to recover as quickly and easily as possible, which might take the form of leaving a river floodplain uninhabited and using it only for seasonal pasture, as was done along the Rhine until engineers straightened and narrowed its channel beginning in 1817.

Societies built resistance to nature's shocks as a conscious enterprise. In regions of the world prone to drought, they developed water-storage infrastructure such as cisterns. In flood-prone regions, they built levees. Cities developed quarantine routines to try to prevent epidemics. By the eighteenth century, China's Qing dynasty had constructed an elaborate system of state granaries intended to prevent famine from whatever cause. (The

Aztecs had done this on a smaller scale in the fifteenth century.) By the nineteenth century, richer societies undertook to control river floods with dikes, dams, and canalization, as on the Rhine, for example.[10] Since the 1880s public health services have made major efforts—by and large crowned by success—to prevent epidemics, by means of sanitation reforms and vaccination regimes. Otherwise there would not be 6.4 billion people today.

There have always been limits to the degree to which resistance can be built. Preventing volcanic eruptions remains impossible and stopping lava flows implausibly expensive. Flood control is feasible but only within limits; levees and dikes occasionally are overwhelmed, as occurred in the Mississippi basin in 1927 and 1993 and most recently in New Orleans in 2005. Moreover, as the Mississippi and New Orleans floods show, societal faith in the infrastructure of resistance can undermine resilience: the opportunity cost of leaving a floodplain unoccupied seems excessive if one trusts the levees and dikes.

Resilience, on the other hand, has to date proved to be in abundant supply: our species has survived countless shocks and now covers the globe as never before. In our earliest years, resilience consisted mainly of mobility—the ability to escape the worst of a natural shock through migration and to start afresh in a new landscape. Until recent decades, this remained an option for millions of pastoralists and the few remaining hunting and foraging populations. As recently as 1912 to 1915, when severe droughts affected the West African Sahel, millions of people adapted by migrating southward—a feasible response because in those days West Africa had about one-eighth the population it carries today, and there were no effective border control regimes to inhibit migration. For the great majority of our historical experience, mobility was the survival response to nature's shocks. Today it is severely restricted.

A second source of resilience in times past was simplicity combined with fertility. Societies with minimal infrastructure lost little except people when a natural disaster struck, and new people were easily created. Rebuilding a complex city in the aftermath of a flood or earthquake requires much more knowledge, investment, coordination, and cooperation than does rebuilding a patchwork of fields and villages. Most peasant societies prior to the twentieth century had a large number of unmarried young people who, in the wake of deadly catastrophe, would stampede into marriage and within a year sharply raise birth rates. This was not a conscious strategy, but a result of custom and economic preferences. Nonetheless, it provided resilience in the form of the ability to ramp up fertility quickly and jump-start demographic recovery.[11]

For many centuries societies have also consciously created more mechanisms to improve resilience. Storing food in state warehouses to cope with famine is a strategy intermittently practiced since ancient times and brought to a high level of reliability by the Qing dynasty in eighteenth-century China.[12] Transportation infrastructure, although built for other reasons, also provided resilience in that it allowed both faster evacuations from affected zones and quicker rescue and relief. Thus societies with extensive and dense road or canal networks, or both, eliminated famine by the end of the eighteenth century, while those lacking transportation infrastructure remained vulnerable.

Organized relief efforts also improved resilience in modern history. The practice of maintaining contingency funds against disasters is probably nearly as old as money and treasuries. The practice by governments of providing funds for disaster victims in other countries dates back at least to a great Jamaican hurricane of 1783 and a Venezuelan earthquake of 1812. The first standing international body devoted to disaster relief probably was the Red Cross, founded in 1863, though until the late 1940s it concerned itself almost entirely with victims of war, rather than nature's shocks.[13] The total effect of such efforts and organizations upon societal resilience has to date been modest, but they have eased the suffering of millions.

In the last two or three centuries, as societies have grown more complex and as mobility has become less feasible as a societal response, resistance and resilience have come to take more technological and bureaucratic forms, such as granaries, seawalls, and international relief organizations. Since 1950 or so, the ability to evacuate millions and to bring large quantities of food and other supplies, quickly and over great distances, has improved immensely. As a result, the causes of modern famines have typically been war and totalitarian politics, rather than environmental factors.[14] Ironically, the logistical capacity to do such things was in large part developed to meet the military requirements of global war, especially in World War II.

As a consequence of this technological and organizational progress, disease, droughts, floods, and earthquakes that a century or more ago might have killed millions more recently would only kill thousands. This extraordinary ability to mitigate disaster has hinged on the relative stability of international politics since 1945, which has provided an opportunity for what we might call "regimes of resilience" to develop. However, the rapid population growth in these same decades imperils resilience by making it harder to maintain uninhabited or sparsely populated buffer zones, wetlands, mangrove forests, floodplains, and so forth. Resilience in the face of drought or similar shocks can be harder to maintain in more crowded circumstances, as can resistance to infectious disease.

In the past, vulnerability to shock had several components. First and most obvious, the intensity and duration of natural shocks often made all the difference between survival and catastrophe. Societies that could withstand one drought a year with only hunger could not withstand two without starvation. Second, some societies had, by design or accident, less in the way of buffers or resilience than others. For example, a society that had few or no domestic animals could not survive a harvest failure as reliably as could a society that could eat its animals one by one if circumstances required it. Societies that had poor transportation infrastructure could not import food as readily or cheaply as could those with good roads, canals, or, eventually, railroads. Nor could the isolated receive any available government or charitable assistance as easily. Societies that used nearly every available acre as farmland and preserved very little in the way of woodlands or wetlands, such as that of early-twentieth-century rural China, proved more vulnerable to floods than did those that (by accident or design) kept land in reserve, because floodwaters sitting on unpopulated or uncultivated land were merely an inconvenience, not a catastrophe. Societies without active and able public health systems suffered more from epidemics than did those that had such systems.

Less obvious, perhaps, were differences in levels of ecological ignorance. Populations that have lived in one environment for several generations gradually acquire, and usually take pains to transmit, knowledge of how to survive and prosper within the limits of their environment. They also gradually form a sense of the boundary conditions to be expected and know from oral tradition that they must be prepared for adversities—locust invasions, prolonged drought, and so forth—beyond their own personal experience. Populations present for dozens of generations normally had exquisitely fine-tuned ecological knowledge and knew where to find edible plants to see them through famine, where to find underground water when there was none on the land's surface, and so forth. Such knowledge contributed materially to resilience.

Conversely, in many instances, especially in the last two centuries, the prevalence of cheap transportation and more long-distance migration has meant that many populations found themselves operating experimentally in new environments. This was true of the British and Irish settlers in Australia after 1788, who inevitably misunderstood antipodean ecology and often paid a price for it.[15] It was true of the American farmers on the southern plains, almost all of whom came from more humid climes, who during the 1930s drought naturally presumed that the wetter years of 1915 to 1930 were normal. They were ignorant of the cyclic drought patterns of the plains; through their farming practices they inadvertently turned the southern plains into the Dust Bowl in a routine drought. Ecological ignorance also lay behind the failures of the Soviet Virgin Lands scheme of the 1950s, in which Premier Nikita Khrushchev ordered an area of dry Siberian and Kazakh steppe land the size of California to be planted in wheat, only to see the region experience, within a few years, disastrous drought, dust storms, and harvest failure.

Societal and Political Reverberations

Even though natural shocks regularly took a significant demographic toll, it is worth emphasizing that the great majority of floods, drought, epidemics, and so on had only local or regional effects and took the lives of small numbers of people. In the distant past, because the human population was small, the numbers of victims were small. It has remained true over the past fifty years partly because of luck (nothing really bad has come up since the influenza pandemic of 1918 to 1919) and partly because public health systems, disaster management systems, and so forth have grown remarkably effective. The worst historical era for demographic losses from natural shocks came between 1300 and 1920.

Interestingly, heightened mortality was not the only source of demographic decline connected to natural shocks. When young people's expectations for the future were lowered and their faith shaken, they tended to postpone marriage, either of their own will or because their elders required it. Moreover, married people, in such dark times, found ways to restrict their fertility. Consequently, for the duration of most disasters, and in the wake of those that were especially disheartening, not only did more people than usual die, but fewer than usual were born. Wars and severe economic depressions also produced this effect. Its magnitude varied tremendously, with the degree of discouragement and the availability of knowledge and means for contraception.

Normally, if disaster was followed by good fortunes, exuberant fertility made up for the losses within a few years. In some cases, however, reproductive slowdowns and strikes lasted decades. This appears to have been the case with the native populations of the Americas during and after the relentless epidemics of the sixteenth and seventeenth centuries.

The economic effects of natural shocks, unlike the demographic ones, have tended to grow and grow. But that is mainly for cheerful reasons: the world economy is now so large that there is much more at risk. Global GNP grew fifteenfold in the twentieth century, and more than fourfold in per capita terms.[16] The direct effects of damage to property depended on where disasters occurred. None were worse, in monetary terms, than the Kobe earthquake of 1995, whose costs may have topped $200 billion, and 2005's Hurricane Katrina, whose costs are put variously between $25 billion and $100 billion. The Indian Ocean tsunami of 2004 led to about $10 billion in direct economic losses.

The Kobe earthquake mangled a densely populated and built-up part of Japan, the country's industrial heartland. It killed 4,571 people and knocked down more than 67,000 buildings. The monetary costs came to about 2.5 percent of Japan's 1995 GNP, and led to the failure of financial institutions such as Barings Bank that were deeply invested in the Japanese property market (Japanese property often carried no earthquake insurance).[17]

Whereas storms and earthquakes often had locally devastating economic effects, droughts—at least as measured by government efforts at compensation—by and large did not. In the United States, estimated annual federal expenditures to mitigate the consequences of droughts averaged half a billion dollars between 1953 and 1988. Federal costs rose from the 1950s to the 1980s, but even the worst case, the drought years from 1987 to 1989 on the Plains, did not much exceed $2 billion a year. During the Dust Bowl decade of the 1930s the government provided far less.[18]

Discrete natural shocks such as hurricanes or floods proved more costly. In the 1950s the American total economic losses came to roughly $4 billion per annum on average. By 2003 that figure had swollen to $65 billion, and in 2004 to $145 billion, according to Munich Re, the world's biggest reinsurance firm. About two-thirds of the costs incurred came from floods and storms. The mass migration into flood-prone regions since 1930 and the consequent creation of housing stock and infrastructure chiefly account for the tremendous rise in the cost of floods and storms. Florida's Broward County, a routine landfall for hurricanes, had 20,000 people in 1930; by 2000 the population was 1.6 million.[19]

Although the costs from nature's shocks rose rapidly—and the shocks could have devastating local effects for a decade or more—none in modern history, not even the 1918-to-1919 influenza, had durable economic consequences that changed the basic fortunes of nations. One could not make that claim for the 1346-to-1350 plague pandemic, which is credited with helping to end feudalism in Western Europe by raising the negotiating power of laborers. But this event, which killed perhaps one-third of Europe's population, was of unique intensity.

A final consideration with respect to the economic implications of nature's shocks is the possibility of "creative destruction," a notion developed by the Austrian economist Joseph Schumpeter. Schumpeter had in mind business cycle crashes and disruptive innovations when he coined this phrase in 1942 to refer to a phenomenon in which bankruptcies eliminated inefficient enterprises, freeing up resources for more efficient use. Taking the response to the plague pandemic in Europe as an inspiration, it is possible to imagine that in the long run, brutal destruction of existing infrastructure and business plant could clear the way for a new generation of more efficient investment. This optimistic perspective, it must be said, assumes that a shock is followed by a time of stability and other favorable conditions. The great Lisbon earthquake of 1755 cleared the way for a more economically rational city plan in subsequent years, but it is anything but clear that post-Katrina New Orleans will feature more economically efficient business plant and infrastructure—although the opportunity surely exists.[20] In any event, recurrent shocks would prohibit creative destruction even if other circumstances were favorable.

Political and social effects of nature's shocks defy quantitative measure, and all conclusions about them are tentative and subject to dispute. Nevertheless, some generalizations seem reliable.

First, nature's shocks in the past have proved simultaneously socially divisive and unifying. This is easily visible in the Katrina disaster. In the wake of the storm, looting was widespread and citizens preyed upon one another in various disturbing ways.[21] Moreover, the challenges of responding to a disaster on that scale exacerbated political and social cleavages, as various officials and groups blamed one another for mismanagement—not without cause. At the same time, however, citizens throughout the United States donated money, materials, and labor in solidarity with the Katrina victims, as did populations in dozens of countries overseas. Such paradoxical responses are probably the norm.

Second, social conflict on some scale was routine during and after disasters. Societies with little in the way of safety net, such as, say, Ethiopia in the 1970s and 1980s, easily succumbed to banditry, ethnic and religious violence, and even outright civil war under the stress of acute drought.[22] Restraint and civility can quickly perish when people are confronted with imperious necessity. This much has been obvious to observers since Thucydides' analysis of the Corcyran Revolution, during the Peloponnesian War.

Third, political reaction to shocks often took the form of scapegoating minorities or foreigners. The Black Death in Europe intensified persecution of Jews, who were accused of poisoning wells and causing the pestilence. This played some role in encouraging Jewish migration to eastern Europe in the fourteenth century.[23] After the great 1923 Kanto earthquake in Japan, which killed some 130,000 to 150,000 people, vigilante mobs together with army and police units attacked Tokyo's Korean community, then about 30,000 strong, and killed perhaps 6,000. Many Japanese believed rumors that Koreans had set fires and poisoned water supplies in the earthquake's aftermath.[24]

Fourth, in the wake of disasters, government authorities have frequently been the target of popular wrath, either for neglect or for intrusive efforts to minimize or prevent damage. This is by and large a modern phenomenon, a reflection of the state's assumption of responsibility for public health and order. Cholera epidemics in nineteenth-century Europe intensified divisions within society and contributed to the revolutionary spirit of the 1830-to-1871 era. Cholera was a fearsome scourge that killed quickly and seemed to come out of nowhere—it is communicated by a bacillus that thrives in warm water and originally came from South Asia. Urban populations with unsanitary water were especially victimized, which in the context of the times fueled the widespread belief that the upper classes or the state were systematically poisoning the poor. Government efforts at quarantines, compulsory hospitalization, and cordons sanitaires provoked riots and attacks on state officials. Even though popular reactions to cholera and to state efforts to control it in France cannot be said to have caused the revolutions of 1830 or 1848, they surely contributed to the distrust of authorities and class antagonisms that underlay these uprisings.[25] Echoes lasted as late as the cholera epidemic in Apulia, Italy, in 1910 and 1911; the authorities reacted by encouraging pogroms against Gypsies (a minority deeply unpopular among other segments of Italian society and with the Italian state) and by forcibly detaining and isolating the sick. Italians responded by persecuting the Gypsies, but also by rioting and killing medical officials, which led the government to call in the army.[26]

In the course of the nineteenth and early twentieth centuries, states took more and more responsibility for public health. Compulsory inoculation against smallpox, pioneered by George Washington in the Continental Army, set an example that inspired much imitation once vaccines were developed against commonplace diseases. (The Revolutionary War might have been lost if Washington had not taken this step.)[27] Yet popular resistance to such measures persisted. In Rio de Janeiro, in 1904 to 1905, poor neighborhoods revolted against public health campaigns involving smallpox vaccination and mosquito control as a measure against yellow fever.[28]

In colonial contexts this sort of political turmoil as a reaction to government efforts to check epidemics or other natural disasters was often still more pronounced, and rumors of deliberate biological warfare against the poor, more frequent. In colonial Mexico, droughts often preceded peasant uprisings, not merely because drought meant hunger but also because at such times the distribution of irrigation water seemed especially unfair, whereas in times of plentiful rainfall it mattered less.[29] In colonial East Africa, efforts to control outbreaks of sleeping sickness that involved resettlement schemes, quarantine of livestock, and other intrusive measures regularly provoked local rebellions against British rule.[30] Along the coast of West Africa, in the area that is now southeastern Ghana, coastal erosion, which the colonial government declined to deal with, helped push the local population into political resistance to colonial rule.[31] British efforts to improve public health in colonial India, and especially to contain the many epidemics of the years 1890 to 1921, frequently ran afoul of local sensibilities and aroused ire that easily translated into political resistance.[32] In the right social and political circumstances, natural shocks, and perceptions of official reactions to them, could precipitate resistance and rebellion.

In one sense, this was nothing new. In most precolonial African societies, and also in imperial China before 1911, populations normally believed that proper ecological functioning—meaning the absence of floods, droughts, epidemics, and so forth—depended on a proper relationship between their rulers and heavenly powers. Natural shocks, therefore, represented a breakdown in that relationship and an inevitable loss of moral authority for rulers. Floods and droughts were taken to mean rulers had lost their efficacy—had lost the "mandate of heaven," in Chinese parlance—and thus no longer were owed obedience. This obviously invited political turmoil.

In the nineteenth and twentieth centuries, when national governments increasingly sought and took responsibility for disease control, flood control, drought relief, and so forth, they inadvertently put themselves in the vulnerable position of the Chinese emperors. If natural shocks were not properly managed or, in some instances, if they were not prevented, the blame lay with the state. Legitimacy became hostage to the whims of nature. So while states improved their capacity to deal with nature's shocks, they were held to ever higher standards, expected to cope effectively with them, but not to intrude too deeply upon citizens' lives and lifestyles. At times rulers invited trouble by encouraging lofty expectations. In 1857, when Napoleon III, the emperor of France, addressed parliament, he had the great Alpine floods of 1856 as well as the revolutions of 1848 on his mind: "By my honor, I promise that rivers, like revolution, will return to their beds and remain unable to rise during my reign."[33] Such boasts did nothing to enhance his moral authority.

The political significance of nature's shocks normally played out on a local or national scale and touched international politics only indirectly. When they did affect international politics, they exhibited the same paradoxical power to bring nations together and to push them into conflict.

Natural disasters have occasionally provoked outpourings of sympathy, both among populations and among states, since at least the eighteenth century. A notable recent example came in August and September 1999, when earthquakes hit first Izmit, in Turkey, and then a suburb of Athens. The Greek government was the first to come to the aid of Turkish earthquake victims, and weeks later the Turks reciprocated. Ordinary Greeks and Turks donated money and supplies to help earthquake victims in the other country. This occurred against a backdrop of long enmity between the governments and populations and helped considerably in defusing a long-simmering rivalry and reorienting politics across the Aegean. Of course, political conditions had to be favorable for a rapprochement before earthquake diplomacy could yield such results: in an era of saber rattling, the governments involved would not have cooperated and would have prevented rather than encouraged generosity on the part of ordinary Greeks and Turks.

Epidemics, while providing plenty of opportunity for mutual recrimination, probably brought states together more often than they drove them apart. The obvious rewards to international cooperation in disease control put the incentives clearly in favor of harmonized actions wherever possible, and against giving vent to frustrations with inadequate measures taken by neighboring states. Since the establishment of the International Red Cross, the World Health Organization, and other such global and regional entities, the multinational integration of disease control efforts has become routine and rarely has been the occasion for conflict. One partial exception to this rule is the position taken by Thabo Mbeki and some other South Africans on HIV/AIDS, which they sometimes attributed to malevolence on the part of Americans and Europeans.[34] Even this, however, did not fundamentally affect relations between South Africa and the West.

Sometimes, of course, nature's shocks exacerbate international or intersocietal conflicts. Earthquakes, hurricanes, and volcanic eruptions have rarely, if ever, had this effect because they are so localized in their damage.

Droughts are another matter. The greatest revolt in the history of Spanish America, that of Tupac Amaru in the Andes from 1780 to 1782, coincided with one of the worst droughts of the millennium, a result of the powerful recurring El Niño current. Thousands of desperate peasants rallied to his standard, which in better times would have appealed to far fewer. In another dramatic case, drought in southern Africa in the decade between 1820 and 1830 converted routine competition for grazing land and food into systemic conquests of the weak by the strong. The *mfecane* ("crushing") created a torrent of refugees throughout southern Africa and resulted in the formation of powerful new states, such as the Zulu kingdom.[35] Drought was also a spur to the slave-raiding that fed the Atlantic slave trade between 1550 and 1850: when food was scarce, one of the few ways to get it was to capture people and trade them for food from afar. Indeed, progressive desiccation—secular climate change in the West African Sahel—drove mounted slave raiders to penetrate deeper and deeper into West Africa in the years after 1600.[36]

Throughout the semiarid zones of the world, where drought was a regular risk, pastoralists and cultivators often uneasily shared frontier zones. Droughts, plagues of locusts, or any natural shock created desperation and drove otherwise peaceful communities to attack their neighbors in hopes of taking their food, livestock, or marketable possessions; and weakness born of drought or some other shock aroused the cupidity of nearby peoples or states suddenly presented with an easy opportunity to raid neighboring communities. The most common format for such violence was attacks by pastoralists upon settled villages, a common pattern in world history in semiarid areas from Manchuria to Senegal. Such attacks of course also took place without the provocation of drought, but drought made them more frequent. In medieval times in northern Syria and Iraq environmental shocks of one sort or another came once every five or six years on average, and often brought political violence in their wake. Villagers had every reason to support a strong state in hopes of keeping marauding pastoralists in check.[37]

While drought was probably the most politically dangerous of all nature's shocks in the deeper past, in the last hundred years water management schemes have often blunted its impact. Moreover, violent political conflict has become more often the affair of states with large urbanized populations rather than pastoral tribes and confederacies, and such states have found it imprudent to go to war to resolve problems created by drought. Even the potentially divisive cases of international river basins such as the Indus, the Mekong, or the Nile have so far been the subject of successful diplomacy rather than military conflict. Observers in recent decades have often foreseen "water wars," in these and other contexts, but it has yet to happen, and indeed it has not happened for several millennia, if ever.[38] The historical record suggests that with well-organized states, the probability of warfare arising from drought-induced water shortage is low; the risk rises in the presence of weak states within which those components of society most aggrieved by drought are less constrained in their responses.

Before departing the subject of political reverberations from nature's shocks it is worth considering whether or not there is an analogue to Schumpeterian creative destruction in the political realm. Can natural shocks shake a society and state out of harmful complacencies and create the political will to undertake needed reforms? Can they discredit the least efficient parts of the political apparatus so thoroughly as to create new space for the more efficient? Perhaps, if conditions already exist for reform and if the gales of destruction are not so powerful as to destroy the state entirely. In the United States, for example, the Dust Bowl of the 1930s gave rise to a useful reform in the creation of the Soil Conservation Service, which has helped prevent the recurrence of catastrophic erosion on the scale of the 1930s, despite droughts in subsequent decades that were equally or more severe. The 1755 earthquake in Lisbon provided the Marques de Pombal with an opportunity to push through fundamental reforms in Portugal. The bubonic plague that harrowed Russia in the 1770s and the cholera epidemics of nineteenth-century Europe both led to major reform efforts in municipal and national governments. Disappointing responses to hurricanes in nineteenth-century Cuba had similar effects.[39] This may amount to a small silver lining in the dark cloud of natural disaster, in the same way that losing a war or undergoing economic depression has served as a spur for reform—provided something survived to be reformed.

Religious turbulence has long been a normal social reaction to nature's shocks. Throughout history most people understood plagues, hurricanes, droughts, and other disasters as divinely ordained or the work of evil people with supernatural powers. Hence, extraordinary natural shocks often brought heightened religiosity, either in the form of more intense devotion to traditional religions or more defections to innovative religions or cults. The rise of the Lotus Sect (Nichiren Buddhism) in Japan was abetted by a great earthquake in Kamakura, one of Japan's chief Buddhist centers, in 1257. The recurrent bubonic plague epidemics in Europe after 1348 gave rise to all manner of eccentric religious practices, most famously a sect of self-flagellants who when not occupied murdering Jews and clergymen wandered about rending their flesh in imitation of Jesus' sufferings. The Neapolitan cult of San Gennaro derives from the experience of 1631, when Naples was spared destruction by a great eruption of Mount Vesuvius. In the United States, the New Madrid earthquakes of 1811 and 1812, following on serious floods in the Ohio and Mississippi basins, helped the prophet Tecumseh, who allegedly had predicted the earthquakes, rally Native Americans to his religious war against the United States. They also prompted many white Americans to experiment with eccentric religious doctrines.[40] The severe drought of 1991 and 1992 in Zimbabwe, often called the worst in living memory, gave rise to at least three charismatic religious movements as Zimbabweans found divine explanations for their misfortunes more satisfying than hypotheses about perturbations in the atmosphere's Intertropical Convergence Zone.[41]

There is rarely a shortage of people charismatic and persuasive enough to make a convincing case—to those ready to be convinced—that any extraordinary event is a sign that radical religious reform is needed. It would be interesting to know whether the Katrina disaster brought an upsurge in religiosity along the Gulf Coast. In any case, if the future holds more serious extreme weather events in store, it seems likely that the most extreme will generate new forms of religion and intensified commitment to old ones.

A Glance at the History of Technological and Social Change

If we are to stop loading the atmosphere with greenhouse gases, we must either find a technological fix or radically reduce energy usage. The history of technological change in modern times is a bit like the punctuated-equilibrium model of evolutionary biology: there are periods of minimal change interspersed with moments of torrential change. Changes tend to come in clusters. The reason for this is that new technologies, to be widely adopted, must fit in not only with existing and emerging technologies but also with existing and emerging institutional, political, economic, and social frameworks—the "software" of society. Technologies coevolve both with other technologies and with this "software."

This is an encouraging perspective in the sense that it means sudden change might occur at any point. It is discouraging in the sense that it is hard to precipitate technological change; it comes when conditions are conducive, but not until then. The trick is to make conditions more conducive: to use policy to alter the "software" and tweak the technological hardware of society so as to speed the probability of new technologies' acceptance in the arenas of energy use and carbon sequestration.

Indeed, any steps to raise the tempo of technological change would likely be helpful on the climate change front. That is because in times past new technologies were adopted (selected, one might say, to use a Darwinian term) for many things, but not for being environmentally helpful—at least, not until very recently. Today, however, and in the foreseeable future, new technologies are unlikely to be adopted if they are environmentally malign; there is, for the first time in history, a "green filter" that skews the process of technological change. This filter exists partly as a matter of regulatory policy, but partly as an ideological force felt in most, if not all, spheres of society. It is likely to be durable, unless the problems of climate change, and of the environment generally, somehow are resolved. It is well to remember, however, that the process by which one technological cluster (say, that of oil, automobiles, and plastics) replaces an earlier one (iron, coal, railroads) is so complicated that it is impossible to produce on schedule or on demand. Further complicating things, a new cluster may take shape quickly in one location, but the speed of its global spread is another matter.[42]

If technology does not come to the atmosphere's rescue and our own, might social change do the trick? This would require changes in behavior that ratchet down energy use, or at least fossil fuel use. History is deeply discouraging on this front. There are very few examples anywhere of societies (as opposed to hermits and monks) that voluntarily renounced the fruits of high-energy society, or embraced a lower standard of living, as lower energy use, absent gains in efficiency, implies. One might claim that the early Christians embraced poverty, but they were a small minority within Roman society, and the great majority of them were poor to begin with.

Sometimes the abolition of slavery is offered as an example of an altruistic social movement that put moral concerns ahead of economic self-interest. Abolition of the slave trade within the British Empire took about thirty years to achieve; abolition of slavery nearly sixty. It took time to build the political coalitions necessary to overcome the self-interest of Caribbean slaveowners, an entrenched lobby well represented in London's Parliament. The effort involved decades of public relations campaigning, undertaken mainly by men of the cloth, as well as routine pork-barrel politics and logrolling, to use the American terms. Within the United States it took longer and took a war; and as a worldwide movement the abolition of slavery took longer still and required the forcible imposition of external values and morals upon African and Arab societies in which slavery had a long tradition of moral justification.

Abolition was indeed a remarkable development: slavery existed for at least 5,000 years and was nearly totally eliminated within 150. But the economic logic of slavery had begun to wane for a number of reasons when abolition gathered momentum as a social movement, and its global success required the moral and military dominance of nineteenth-century Britain and the self-confidence within British society to force abolition upon unwilling societies and cultures—a constellation of circumstances not easily reproduced. What this example really shows is how exceptional, and how contingent upon economic and political circumstances, the abolition of slavery really was. We could wait a long time for the stars to align themselves just right so as to permit a social movement that would lead to reduced energy use.

It remains theoretically possible that international accords might be reached that would limit the emission of greenhouse gases, as was done with chlorofluorocarbons beginning in 1989. But the odds are against it, for a number of reasons. The ozone accords were low-hanging fruit. Technological alternatives proved easy to find and were manufactured by the biggest CFC-producing firms. CFCs were useful, but amounted to a tiny part of any economy compared with fossil fuels. Only a handful of countries made them. And atmospheric scientists could make a strong case that ozone depletion formed a direct threat to human health in the form of higher risks of skin cancer. While initially CFC manufacturers denied the truth of the science, this phase lasted only briefly. As the unfruitful negotiations over carbon emissions and climate change since the early 1990s show, limiting greenhouse gases is inherently a tougher diplomatic nut to crack.

The Unprecedented Challenge Ahead

History does not tell us much of anything directly about social and political responses to climate change. Abrupt climate change is too far back in the past, when societies were too different from our own, to shed any light on the matter. More recent climate change, when societies were more closely comparable to ours, was too gradual, usually too slow to be noticed. Even the arrival of warmer centuries from 900 to 1200 C.E., known as the Medieval Optimum, and the cooling of the Little Ice Age went undetected at the time.

Widening the lens to consider natural shocks of several sorts, as this chapter does, is a bit more helpful. Such shocks have been part of the ordinary experience of most generations until very recently. The most serious were epidemics and droughts—both of which climate modelers anticipate will become more likely in a greenhouse world. Resilience in the face of such shocks originally consisted mainly of mobility and simplicity of the way of life but within the last two centuries came to rest more and more on bureaucratic provisions for disaster and technological means to prevent or mitigate it.

The ability to do this well was, and is, very unevenly distributed around the world. Assuming that such shocks become more common in the future, the ability to generate and provide resilience, in whatever form, will become an ever greater force in determining the fates of societies and states. The degree to which international institutions master the necessary skills and execute them reliably and equitably will play an ever larger role in determining the levels of tensions within the international system.

The demographic and economic effects of nature's shocks, while often locally or regionally devastating, normally came to little on the global scale. The plague pandemic of the fourteenth century is the greatest exception. In certain circumstances, sizable shocks could have serious politically disruptive and destabilizing effects within societies and occasionally contributed to conflict between societies. Nature's shocks also routinely nudged along processes of social change, and occasionally more than that, notably in the field of religion. To judge by the record of the past, one should expect a greenhouse world to be a bit more volatile politically: stable nation-state regimes will be harder to build and maintain and internationalism will be subject to strains somewhat greater than would otherwise be the case, but with a significant countercurrent pushing toward greater cooperation in the face of common threats. And one should expect stronger religious enthusiasm than has ordinarily prevailed in modern history.

Conclusion

So can history help us with global warming? Yes and no. Yes in the sense that in the long record of human history there have been certain consistent elements in human beings' myriad responses to environmental disasters. Those elements—intensified conflict, deeper cooperation, political rebelliousness, religious zeal—appear in varying mixes, depending on the character of any given disaster and on social and political frameworks involved. The answer also has to be no, however, given that past disasters occurred on a relatively limited or discrete scale, particularly in recent years. There is no precedent in human history for a global disaster that affects whole societies in multiple ways at many different locations all at once. It is very difficult to predict how the past might inform the present and the future when it comes to climate change as a global phenomenon. But the effects of climate change will play out simultaneously on several scales, and some of its likeliest consequences—enhanced drought and flooding, for example—will in the future, as in the past, be felt locally and regionally rather than globally. Thus, the more one unpacks the concept of climate change into its components, the more the record of the past becomes relevant to imagining the future.

Notes

1. Recent accounts of Holocene climate that are easily accessible to nonspecialists include Eugene Linden, *The Winds of Change: Climate, Weather and the Destruction of Civilization* (New York: Simon & Schuster, 2006); Brian Fagan, *The Long Summer: How Climate Changed Civilization* (New York: Basic Books, 2004). Somewhat more technical are William James Burroughs, *Climate Change in Prehistory: The End of the Reign of Chaos* (Cambridge University Press, 2005); Douglas MacDougall, *Frozen Earth: The Once and Future Story of Ice Ages* (University of California Press, 2004).

2. The best study remains Jean Grove, *The Little Ice Age* (Cambridge University Press, 1988).

3. William H. McNeill, *Plagues and Peoples* (New York: Doubleday, 1976) remains the most useful survey of disease history.

4. On the AIDS epidemic mortality, see Kenneth Mayer and H. S. Pizer, *The AIDS Pandemic: Impact on Science and Society* (New York: Academic Press, 2004), p. 2.

5. For higher figures, see Mike Davis, *Late Victorian Holocausts: El Niño and the Making of the Third World* (London: Verso, 2002).

6. V. K. Sharma and A. D. Kaushik, "Floods in India" (www.nidm.net/Chap6.htm); Xia Jun and Y. D. Chen, "Water Problems and Opportunities in the Hydrological Sciences in China," *Hydrological Sciences* 46, no. 6 (2001): 907–21; Gerd Tetzlaff and others, "Das Jahrtausendhochwasser von 1342 am Main aus meteorologisch-hydrologischer Sicht," *Wasser und Boden* 54 (2002): 41–49.

7. I purposefully omit mention of the flood in Johnstown, Pennsylvania, in 1889, a case of a burst dam, which killed 2,200. This I consider not one of nature's shocks but a case of engineering failure.

8. Jelle Zeilinga de Boer and Donald S. Sanders, *Earthquakes in Human History* (Princeton University Press, 2005). By far the most serious eruption, from the human point of view, took place around 74,000 years ago, at Lake Toba, Sumatra. This eruption ejected 2,800 times as much material as Mount St. Helens, covered Malaysia in up to 9 meters (30 feet) of ash, put 10 billion tons of sulfuric acid into the atmosphere (producing acid rain of a unique intensity), lowered global temperatures by 3 to 4°C (5 to 7°F), and, according to one interpretation of mitochondrial DNA evidence, brought the human race to the brink of extinction. This was the biggest eruption of the past 2 million years.

9. Of course, the rich world has since 1919 lived with the specter of war and since the 1940s with the risk of nuclear annihilation.

10. Mark Cioc, *The Rhine: An Eco-Biography* (University of Washington Press, 2002); David Blackbourn, *The Conquest of Nature: Water, Landscape, and the Making of Modern Germany* (New York: Norton, 2005).

11. Jacques Dupâquier and others, *Marriage and Remarriage in Populations of the Past* (London: Academic Press, 1981).

12. Pierre-Etienne Will and R. Bin Wong, *Nourish the People: The State Civilian Granary System in China, 1650–1850* (University of Michigan Center for Chinese Studies, 1991). When Henry Wallace read of this system in a Columbia University dissertation on Confucian economics of 1911, he drew inspiration that informed his policies as Secretary of Agriculture under Franklin Roosevelt, and helped shape the mission of the UN's Food and Agriculture Organization.

13. Matthew Mulcahy, *Hurricanes and Society in the British Greater Caribbean, 1624–1783* (Johns Hopkins University Press, 2006); Zeilinga de Boer and Sanders, *Earthquakes in Human History,* p. 128; and David P. Forsythe, *The Humanitarians* (Cambridge University Press, 2005).

14. Cormac Ó Gráda, "Making Famine History," *Journal of Economic Literature* 45 (2007): 5–38.

15. Tim Flannery, *The Future Eaters* (New York: Baziller, 1994).

16. J. R. McNeill, *Something New under the Sun* (New York: Norton, 2000), p. 361, based on Angus Maddison, *Monitoring the World Economy* (Paris: OECD, 1995).

17. City of Kobe, "The Great Hanshin-Awaji Earthquake Restoration Project: Statistics and Restoration Progress" (2005) (http://www.greenbar.org/kobestats.htm).

18. Presumably more was spent on drought relief at state and local levels. Western Governors Association, "Western Drought Facts and Information 2004" (www.westgov.org/wga/testim/drought-fact04.pdf). See also W. E. Riebsame, S. A. Changnon, and T. R. Karl, *Drought and Natural Resources Management in the United States: Impacts and Implications of the 1987–89 Drought* (Boulder, Colo.: Westview Press, 1991).

19. Holli Riebeek, "The Rising Cost of Natural Hazards," report, March 28, 2005, NASA's Earth Observatory (earthobservatory.nasa.gov/Study/RisingCost/printall.php).

20. William L. Waugh and R. Brian Smith, "Economic Development and Reconstruction on the Gulf after Katrina," *Economic Development Quarterly* 20 (2006): 211–18.

21. See "Storm and Crisis: Lawlessness," *New York Times,* September 25, 2005, p. A1, which offers an initial assessment of the uptick in assaults, rapes, and murders that took place. The numbers cannot be verified because of the breakdown in policing, in emergency services, and in hospital record keeping.

22. Sven Rubenson, "Environmental Stress and Conflict in Ethiopian History: Looking for Correlations," *Ambio* 20 (1991): 179–82.

23. Catherine Park, "The Black Death," in *The Cambridge World History of Human Disease,* edited by K. N. Kiple (Cambridge University Press, 1993), pp. 612–15.

24. Zeilinga de Boer and Sanders, *Earthquakes in Human History,* pp. 188–89. Details available in Sonia Ryang, "The Great Kanto Earthquake and the Massacre of Koreans in 1923: Notes on Japan's Modern National Sovereignty," *Anthropological Quarterly* 76, no. 4 (2003): 731–48.

25. Louis Chevalier, ed., *Le choléra: la première épidémie du XIX siècle* (La Roche-sur-Yon, 1958); and Richard Evans, "Epidemics and Revolutions: Cholera in Nineteenth-Century Europe," in *Epidemics and Ideas: Essays in the Historical Perception of Pestilence,* edited by Terence Ranger and Paul Slack (Cambridge University Press, 1992): 149–73.

26. Frank Snowden, *Naples in the Time of Cholera, 1884–1911* (Cambridge University Press, 1995).

27. Elizabeth Fenn, *Pox Americana: The Great Smallpox Epidemic of 1775–1782* (New York: Hill & Wang, 2001).

28. N. Sevcenko, *A revolta da vacina: mentes insanas em corpos rebeldes* [The vaccination revolt: insane minds in rebellious bodies] (São Paulo: Scipione, 1993).

29. Georgina Enfield, *Climate and Society in Colonial Mexico: A Study in Vulnerability* (London: Routledge, forthcoming, 2008); D. S. Brenneman, "Climate of Rebellion: The Relationship Between Climate Variability and Indigenous Uprisings in Mid-Eighteenth-Century Sonora," Ph.D. dissertation, University of Arizona, 2004.

30. Kirk Hoppe, *Lords of the Fly: Sleeping Sickness Control in British East Africa, 1900–1960* (Westport, Conn.: Praeger, 2003).

31. Emmanuel Akyeampong, *Between the Sea and the Lagoon: An Ecohistory of the Anlo of Southeastern Ghana* (Ohio University Press, 2001).

32. David Arnold, *Colonizing the Body: State Medicine and Epidemic Disease in Nineteenth-Century India* (University of California Press, 1993).

33. Quoted in Michael Bess, *The Light-Green Society: Ecology and Technological Modernity in France, 1960–2000* (University of Chicago Press, 2003), p. 57.

34. Theodore F. Sheckels, "The Rhetoric of Thabo Mbeki on HIV/AIDS: Strategic Scapegoating?" *Howard Journal of Communications* 15 (2004): 69–82.

35. Elizabeth Eldredge, "Sources of Conflict in Southern Africa, c. 1800–1830: The Mfecane Reconsidered," *Journal of African History* 33 (1992): 1–35.

36. James Webb, *Desert Frontier: Ecological and Economic Change along the Western Sahel, 1600–1850* (University of Wisconsin Press, 1994); Joseph Miller, *The Way of Death: Merchant Capitalism and the Angolan Slave Trade, 1730–1830* (University of Wisconsin Press, 1988).

37. Magnus Widell, "Historical Evidence for Climate Instability and Environmental Catastrophes in Northern Syria and the Jazira: The Chronicle of Michael the Syrian," *Environment and History* 13 (2007): 47–70. The data used here are from 600 to 1200 C.E. The broader pattern was described in the fourteenth century by Ibn Khaldun, *The Muqaddimah* (Princeton University Press, 1967), based on his experiences in North Africa.

38. Sandra Postel and Aaron Wolf, "Dehydrating Conflict," *Foreign Policy* 126 (September–October 2001): 60–67; Thomas Homer-Dixon, "On the Threshold: Environmental Changes as Causes of Acute Conflict," *International Security* 16 (Autumn 1991): 76–116; Thomas Homer-Dixon, "Environmental Scarcities and Violent Conflict," *International Security* 19 (1994): 5–40.

39. John Alexander, *Bubonic Plague in Early Modern Russia: Public Health and Urban Disaster* (Johns Hopkins University Press, 1980); Louis Perez, *Winds of Change: Hurricanes and the Transformation of Nineteenth-Century Cuba* (University of North Carolina Press, 2001).

40. Zeilinga de Boer and Sanders, *Earthquakes in Human History,* pp. 133–35, 190–92; James L. Penick, *The New Madrid Earthquakes* (University of Missouri Press, 1981), pp. 111–26.

41. Hezekiel Mafu, "The 1991–1992 Zimbabwean Drought and Some Religious Reactions," *Journal of Religion in Africa* 25 (1995): 288–308.

42. David Egerton, *The Shock of the Old: Technology in Global History since 1900* (Oxford University Press, 2007).

three

Three Plausible Scenarios of Future Climate Change

JAY GULLEDGE

This chapter reviews projected climate change impacts over the next thirty to one hundred years and outlines three climate change scenarios, of three grades of severity, that cover a plausible range of impact severity. These scenarios, based on current scientific understanding and uncertainty regarding past and future climate change, guide assessments in later chapters of potential security consequences of climate change impacts. The general approach is to settle on three different levels of global average temperature change for each scenario, and then extract relevant projected impacts from the *Fourth Assessment Report* (*AR4*) of the Intergovernmental Panel on Climate Change (IPCC) and other peer-reviewed scientific sources. We focus particularly on changes in freshwater resources, crop production, extreme weather events, sea level rise, and the meridional overturning circulation (MOC) of the North Atlantic Ocean.

Because the purpose of this project is to assess potential security risks of future climate change, the primary criterion for the climate impacts scenarios outlined here is *plausibility* rather than *probability*. Rather than asking, What is the most likely climate-driven outcome? we ask, *What potential climate-driven outcomes are plausible, given current scientific understanding and uncertainties about the future climate?* Recent observations indicate that

The scenarios outlined in this section are not predictions of future conditions and should not be read or cited as such.

projections from climate models have been too conservative; the effects of climate change are unfolding faster and more dramatically than expected. Given the uncertainty in calculating climate change, and the fact that existing estimates may be biased low at this time, plausibility is an important measure of future impacts.

Under this umbrella of plausibility, potential changes that the IPCC or other assessments may characterize as improbable are considered plausible here if significant uncertainty persists regarding their probability; collapse of the North Atlantic MOC is an example. Because projections of sea level rise remain particularly uncertain, direct consultation with experts and the author's professional judgment inform the sea level rise scenarios outlined here. Accordingly, the impacts scenarios presented here are not predictions, but are potential outcomes either supported by or not ruled out by current scientific understanding, and therefore deserving of consideration when potential risks are assessed.

These impacts scenarios are meaningful only in the context of the security risk assessments offered in this volume and should not be misconstrued as predictions. Physically deterministic predictions of future climate are currently impossible, and perfect foresight would obviate in any case the need for scenario-based assessments. This is the unavoidable backdrop of uncertainty against which analysts must assess the implications of global climate change.

Scenario-Based Approach

According to the IPCC, a *scenario* is "a coherent, internally consistent and plausible description of a possible future state of the world. Scenarios are not predictions or forecasts but are alternative images without ascribed likelihoods of how the future might unfold."[1] In this volume we develop a group of three impacts scenarios: expected, severe, and catastrophic. Although guided in general by the IPCC *AR4* and other authoritative scientific sources, these impacts scenarios are unique to this study and were created specifically for its purposes.

The IPCC uses independent scenarios of man-made greenhouse gas emissions called SRES (*Special Report on Emissions Scenarios*) scenarios in its assessment process.[2] The SRES scenarios make assumptions about future population growth, economic and infrastructure development, and energy policy that result in plausible, alternative pathways of future greenhouse gas emissions in the absence of policies to mitigate climate change. In the IPCC assessments and other studies, multiple SRES emissions scenarios are used as plausible alternatives to drive climate models, thus producing multiple plausible projections of future climate conditions. As described later, the SRES A1B emissions scenario is used in our study solely to derive levels of temperature change for each of our three impacts scenarios. We then extract impacts from published studies (primarily the *AR4*) based on those levels of temperature change, regardless of which emissions scenarios were used to drive climate models in those studies.

A caveat of this approach is that different SRES emissions scenarios assume different demographic trends, such as total population, population living near coastlines, and level of economic and technological development in developing countries. These differences alter estimates of population sizes affected by climate impacts, particularly sea level rise, food availability, and water scarcity. To address this caveat, in some cases we present a range of estimates provided in the published literature based on a variety of emissions scenarios for a given temperature change. From the perspective of risk assessment, the upper ends of such ranges are most relevant (see tables 3-2 and 3-3 for more information and examples).

In any assessment of climate change, it is essential to distinguish between a prediction and a projection. A *projection* describes an outcome that is deemed plausible, often subjectively, in the context of current uncertainties, whereas a *prediction* describes the statistically most probable outcome based on the best current knowledge.[3] As described by Michael MacCracken, "A *projection* specifically allows for significant changes in the set of [determinants] that might influence the [future climate], creating 'if this, then that' types of statements."[4] The greater the degree of uncertainty surrounding determinants of future climate conditions, such as future man-made greenhouse gas emissions, the less certain a prediction can be and the more important projections become for risk assessment. This is why the IPCC uses several alternative SRES emissions scenarios in assessing future climate change. In keeping with the purpose of our study, our scenarios outline plausible impacts projections and should not be taken to be or cited as predictions of future conditions.

Underlying Assumptions in the Three Climate Impacts Scenarios

As a basis for outlining future climate change impacts, we derive temperature change projections based on the SRES A1B emissions scenario defined by the IPCC, with upward temperature adjustments for our two more extreme scenarios. SRES A1B is a medium-range emissions scenario that

Box 3-1. Two Myths about Climate Change

Myth 1: Future Climate Change Will Be Smooth and Gradual

The history of climate reveals that climate change occurs in fits and starts, with abrupt and sometimes dramatic changes rather than gradually over time.[1] An example of contemporary abrupt change is evident in the frequency of North Atlantic tropical storms (figure 3-1). This basic feature of abrupt change implies that surprising changes are likely to occur in the future, even if average climate change is projected accurately.[2]

Figure 3-1. Tropical Hurricane Frequency in the North Atlantic[a]

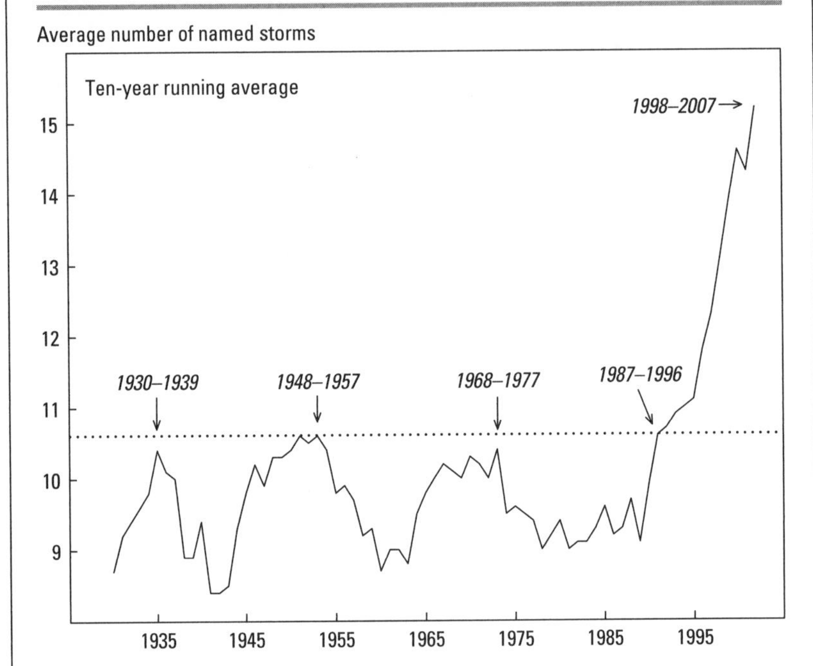

Source: Atlantic Hurricane Database Re-analysis Project (www.aoml.noaa.gove/hrd/data_sub/re_anal .html).

a. The ten-year running average of annual frequency shows a dramatic and abrupt increase, 40 percent, above the previous maximum observed in the mid-1950s, previously considered extreme.

Thus, a projection of 1 meter (3.3 feet) of sea level rise over one century could turn out to be correct, but it could occur in several quick pulses with relatively static periods in between. This type of change is more difficult to prepare for than gradual change, as large-scale public works projects intended to adapt to such a change are likely to require several decades to

Box 3-1. Two Myths about Climate Change (*continued*)

complete. Most climate change scenarios do not incorporate such abrupt changes, as they are based on large-scale averages through space and time. In this volume we consider the possibility that humanity could be surprised by and unprepared for abrupt changes in the severe and extreme climate change scenarios. Surprises from abrupt climate change are likely to increase the burden of climate impacts beyond what is expected, with unforeseen economic and security implications.

Myth 2: Industrialized nations will be spared major impacts.

Many people have the impression that developed nations will not experience serious climate change impacts. In fact, the opposite is likely. The United States, southern Europe, and Australia are likely to be among the most physically impacted regions. By virtue of its large size and varied geography, the United States already experiences a wide range of severe climate-related impacts, including droughts, heat waves, flash floods, and hurricanes, all of which have been or are likely to be exacerbated by climate change.[3] For instance, from 1980 to 2000 the United States was among the countries experiencing large numbers of deaths from exposure to tropical storms (figure 3-2).

According to the IPCC, the western United States, southern Europe, and southern Australia will experience progressively more severe and persistent droughts, heat waves, and wildfires in future decades as a result of climate change.[4] In 2003 central and southern Europe experienced a prolonged heat wave that was the most severe in at least 500 years and led to the premature deaths of 50,000, mostly elderly people.[5] Model projections suggest this type of heat event will become typical for the region by 2040 as a result of the current global warming trend.[6] The United States, with the largest number of coastal cities and two agricultural river deltas near or below sea level, is one of the most susceptible countries to future sea level rise. The United States and coastal countries of the European Union are likely to experience some of the greatest losses of coastal wetlands.[7]

The misapprehension that climate change impacts will spare the industrialized world may stem from confusion between the concepts of *impact* and *vulnerability.* Vulnerability measures the ability of a population to withstand impact, but low vulnerability does not imply low impact. Because it possesses greater infrastructure and wealth, the United States may be more capable of devoting resources to preparing for, adapting to, and recovering from climate change impacts than developing countries with similar exposure. Because it will be severely impacted, the United States will need to divert great financial

(continued)

Box 3-1. Two Myths about Climate Change (*continued*)

Figure 3-2. Relative Vulnerability to Tropical Cyclones, 1980–2000[a]

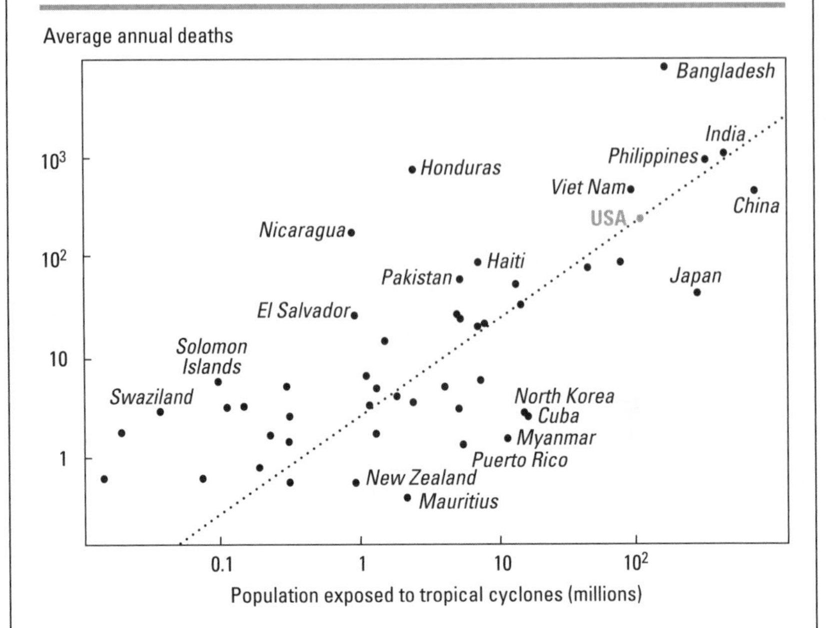

Source: N. Pelling, A. Maskrey, P. Ruiz, and L. Hall, *Reducing Disaster Risk: A Challenge for Development, A Global Report* (New York: United Nations Development Program, Bureau for Crisis Prevention and Recovery, 2004 [www.undp.org/bcpr/we_do/global_report_disaster.shtml]).

a. Countries near the diagonal line, such as India and the Philippines, exhibited vulnerability similar to the United States. Countries below the diagonal line, including China, Cuba, and Japan, were less vulnerable than the United States.

and material resources toward coping with climate change. In general, severe climate change impacts affecting wealthy nations portend diversion of foreign aid to domestic projects, generating greater potential for environmental refugees to attempt to migrate to wealthy countries.

1. Committee on Abrupt Climate Change, *Abrupt Climate Change: Inevitable Surprises* (Washington: National Academy Press, 2002).

2. Ibid.

3. IPCC, "Summary for Policymakers," in *Climate Change 2007: Impacts, Adaptation and Vulnerability;* IPCC, "Summary for Policymakers," in *Climate Change 2007: The Physical Science Basis; Contribution of Working Group I to the Fourth Assessment Report of the Intergovernmental Panel on Climate Change,* edited by Susan Solomon and others (Cambridge University Press, 2007), pp. 1–18; 7–22.

4. Ibid.

5. J. Alcamo and others, "Europe," in *Climate Change 2007: Impacts, Vulnerability and Sustainability: Contribution of Working Group II to the Fourth Assessment Report of the Intergovernmental Panel on Climate Change,* edited by M. L. Parry and others (Cambridge University Press, 2007).

6. P. A. Stott, D. A. Stone, and M. R. Allen, "Human Contribution to the European Heatwave of 2003," *Nature* 432 (2004): 610–14.

7. R. J. Nicholls, "Coastal Flooding and Wetland Loss in the 21st Century: Changes under the SRES Climate and Socio-Economic Scenario," *Global Environmental Change* 14 (2004): 69–86; R. J. Nicholls and others, "Coastal Systems and Low-Lying Areas," in *Climate Change 2007: Impacts, Adaptation and Vulnerability,* pp. 315–56.

considers continued growth of man-made greenhouse gas emissions under rapid economic growth, technological development, and ongoing efficiency improvements, but with significant continued reliance on fossil fuels.[5] Atmospheric carbon dioxide (CO_2) rises to a concentration of about 700 parts per million (ppm)—2.5 times the preindustrial concentration—by the end of the twenty-first century, which the *AR4* projects would be associated with a global surface temperature increase of 1.7 to 4.4°C (3.1 to 7.9°F), with a best estimate of 2.8°C (5.0°F).[6] Although SRES scenarios assume that society takes no actions to limit climate change, it is possible for society to enact policies that would limit emissions significantly below the level of the A1B projection.[7]

An unavoidable caveat of basing our impacts scenarios on IPCC model projections is that the regional projections are continental or subcontinental in scale and impacts are generally described in aggregate. How climate in any specific location might deviate from the subcontinental average is uncertain; it might not be possible to predict correctly the consequences of climate change for particular locales from the existing scientific literature. As a result, assessing the security implications of climate change requires making some assumptions regarding the impacts that may occur in a given geopolitical arena.

Two of the impacts scenarios outlined here project changes to the year 2040. Although we choose a particular emissions scenario as a reference case, temperature increases based on the various emissions scenarios examined by the IPCC do not diverge significantly by the year 2040, as past emissions dominate temperature forcing over this short time frame. Uncertainty in the temperature outcome within this time frame is related less to greenhouse gas emissions than to uncertainty about physical climate sensitivity to greenhouse gas forcing and the response of individual climate components (for example, ice sheets, sea level, or storm systems) to a given degree of warming.[8] Over the longer time frame (about one century) of the most severe scenario, divergence of different emissions scenarios is significant and A1B emerges as a mid-range projection of temperature change, which we adjust in scenario 3 to account for potential underestimation as described below.

Climate Scenario 1: Expected Climate Change

This scenario provides the basis for chapter 4 in this volume, by John Podesta and Peter Ogden, on the *expected* consequences of climate change for national and international security over the next thirty years. It accepts the temperature change projected in the *AR4* for emissions scenario A1B,

Table 3-1. Global Average Surface Warming and Sea Level Rise Relative to 1990 for the Plausible Scenarios of Climate Change

Climate scenario	Start year	End year	Warming	Basis for warming	Sea level rise
Scenario 1 (expected)	1990	2040	1.3°C (2.3°F)	Model average for A1B emission scenario in 2040	0.23 m (0.75 ft)
Scenario 2 (severe)[a]	1990	2040	2.6°C (4.7°F)	Double the model average for A1B in 2040	0.52 m (1.7 ft)
Scenario 3 (catastrophic)[a]	1990	2100	5.6°C (10.1°F)	Double the model average for A1B in 2100	2.00 m (6.6 ft)

Source: Author's compilation.

a. Projections for scenarios 2 and 3 are unique to this study and are meaningful only within the context of this study.

resulting in warming of 1.3°C (2.3°F)) relative to 1990 (table 3-1). Attendant impacts described for this temperature change are also accepted, except for sea level rise, which is assessed separately as described below. The *AR4* projects impacts for the 2020s, 2050s, and 2080s. Where relevant, scenario 1 assumes that impacts intermediate to those described for the 2020s and 2050s represent impacts thirty years from the present.

Climate Scenario 2: Severe Climate Change

This scenario provides the basis for chapter 5 in this volume, by Leon Fuerth, on *severe* consequences of climate change for national and international security over the next thirty years. It assumes that the *AR4* projections of both warming and attendant impacts are systematically biased low. Multiple lines of evidence support this assumption, and it is therefore important to consider from a risk perspective.[9] For instance, the models used to project future warming either omit or do not account for uncertainty in potentially important positive feedbacks that could amplify warming (for example, release of greenhouse gases from thawing permafrost, reduced ocean and terrestrial CO_2 removal from the atmosphere), and there is some evidence that such feedbacks may already be occurring in response to the present warming trend.[10] Hence, climate models may underestimate the degree of warming from a given amount of greenhouse gases emitted to the atmosphere by human activities alone. Additionally, recent observations of climate system responses to warming (for example, changes in global ice cover, sea level rise, tropical storm activity, average annual precipitation) suggest that IPCC models underestimate the responsiveness of some aspects of the climate system to a given amount of warming.[11] On these premises, the second scenario assumes that omitted positive feedbacks occur quickly and amplify warming to double that projected for emissions scenario SRES A1B, and that the climate system components respond more strongly to warming than predicted (see table 3-1).

According to our current understanding of physical inertia in the climate system, warming of 2.6°C (4.5°F) seems highly unlikely on the thirty-year time scale. Bearing in mind, however, that the IPCC projections show only average change with a smooth evolution over time and have tended to underestimate climate system response to warming already realized, a combination of underestimated change and abrupt episodes could plausibly result in an unexpectedly large and rapid warming with larger than expected impacts in a matter of a few decades. Moreover, a recent study aimed at quantifying the uncertainty surrounding model projections of future temperature found a greater than one-in-twenty chance that by 2040, warming could exceed 2°C (3.6°F) relative to 1990 for the highest SRES emissions scenario in the absence of strong positive feedbacks and abrupt change.[12]

Climate Scenario 3: Catastrophic Climate Change

This scenario provides the basis for chapter 6 in this volume, by Sharon E. Burke, on *catastrophic* consequences of climate change for national and international security through the end of the twenty-first century. On the basis of current scientific understanding, we assume that abrupt, global catastrophic climate events cannot plausibly occur in the next three decades, but could plausibly do so over the course of this century. To examine the consequences of such events, scenario 3 extends the rapid warming and attendant accelerated impacts associated with scenario 2 to the end of the twenty-first century, leading to assumed rapid loss of polar land ice, abrupt 2-meter (6.6-foot) sea level rise, and the collapse of the North Atlantic MOC. We therefore assume warming that is double the best estimate of modeled surface warming under emissions scenario A1B for the year 2100 (see table 3-1).

Although doubling an IPCC projection is arbitrary, the result—5.6°C (10.1°F) warming by 2095 relative to 1990—compares well with the upper-end projection of a group of models that incorporated carbon cycle feedbacks and therefore simulated higher atmospheric CO_2 growth rates than did the IPCC models.[13] When adjusted to account for changes in non-CO_2 greenhouse gases and atmospheric particulates, the models including carbon cycle feedbacks produced an upper-end projection of a 5.6°C (10.1°F) temperature rise in 2100 relative to 2000. These models still did not incorporate all possible positive feedbacks, such as increased greenhouse gas emissions from thawing permafrost, so our most extreme warming scenario could potentially prove conservative. Even so, there is little utility in assuming higher projected temperatures, as impacts have generally not been assessed for twenty-first-century warming greater than 5°C (9°F).[14]

Sea Level Rise

Given that 10 percent of the world's population currently lives in low-lying coastal zones and that this proportion is growing, sea level rise is an important aspect of future climate change impacts.[15] Unfortunately, current methods of projecting sea level are insufficient to provide either a best estimate or an upper limit for sea level rise over the current century.[16] This section describes how plausible values for average global sea level rise were derived for the three climate change scenarios (see table 3-1).

Uncertainties Regarding Future Sea Level Rise

Glaciers and ice sheets form streams of ice that flow in solid form from land into the ocean. If the flow rate of these ice streams accelerates without an equivalent increase in snow deposition on the glacier or ice sheet, sea level rise will accelerate. The range of sea level rise projected for the twenty-first century in the *AR4* explicitly omits any estimate of future ice flow acceleration in response to global warming into the ocean from the Greenland and Antarctic ice sheets, yet recent observations indicate that ice flow is already accelerating on parts of these ice sheets.[17] IPCC sea level projections also assume that melt ponds on the surface of ice sheets refreeze on the ice sheet rather than draining to the ocean, whereas recent observations and theoretical assessment suggest that an unknown fraction of this meltwater finds its way into the ocean.[18] These ice sheets represent the largest potential source of future sea level rise, and omitting ice sheet dynamics and melt pond drainage likely systematically biases the IPCC projections low. For the IPCC, this omission was perhaps unavoidable because current knowledge of ice sheet dynamics simply does not permit the process to be modeled. For our purposes, such an omission is unacceptable as it would lead to an unrealistically low upper limit. We therefore depart from the *AR4* to assess plausible upper limits to sea level rise under our two most severe scenarios.

The IPCC's model projections for sea level rise from the 2001 *Third Assessment Report* (*TAR*) were higher than the latest projections of the *AR4*.[19]

A recent study demonstrated that observed sea level rise for the period 1990 to 2006 tracks the upper uncertainty bound of the TAR projections, and therefore exceeds all IPCC model projections for sea level rise during the same period.[20] For scenario 1, therefore, we adopt the upper bound of projected sea level rise in the *TAR*. This approach yields a sea level rise for scenario 1 of 23 cm (9 inches) in the year 2040 relative to 1990. (Note that all IPCC scenarios of twenty-first-century sea level rise are relative to global average sea level in 1990.)

The greatest uncertainty regarding future sea level rise is whether it will be dominated by thermal ocean-water expansion or by contributions of freshwater from land-based ice sheets; only the latter could generate abrupt sea level rise of 1 meter (3.3 feet) or more by the end of this century. The IPCC stated clearly in the *AR4* that it could provide neither a best estimate nor a plausible upper end for sea level rise in the event that future sea level rise is dominated by accelerating ice sheet contributions.[21] For scenarios 2 and 3, therefore, the author surveyed nine leading climatologists with relevant expertise to assess to what extent and by what means twenty-first-century sea level rise can be constrained at the upper end.[22] This is an accepted approach for assessing climate change when fundamental uncertainties hamper model-based estimates.[23]

All of the experts agreed that at least 1 meter (3.3 feet) of sea-level rise by the end of the twenty-first century was plausible, and two experts viewed 1 meter as a plausible lower limit. Several declined to constrain the upper end, citing the *AR4* position that current knowledge and methods are insufficient to allow it. Three experts cited the observation-based projection of the ocean physicist Stefan Rahmstorf as providing a plausible upper limit of 1.4 meters (4.6 feet) for the end of the twenty-first century.[24] One of those experts also felt that higher projections could not be justified. One expert presented 2 meters (6.6 feet) as a firm upper limit, and another said that 2 meters "or even more" was plausible.

In recent peer-reviewed publications, Rahmstorf has concluded that more than 1 meter (3.3 feet) of sea level rise could not be ruled out, and the climate physicist James Hansen expressed confidence that sea level rise would be measured in meters on the century time scale.[25] Based on assumed logarithmic acceleration of land-based ice loss, Hansen infers that an ice sheet contribution on the order of 5 meters (16.4 feet) during the twenty-first century is plausible. One of the experts interviewed doubted that Hansen's analysis could be defended adequately on the basis of current scientific understanding (others did not comment on Hansen's assessment), although the same expert felt that it was currently impossible to constrain twenty-first-century sea level rise to within less than a tenfold range of uncertainty and therefore conceded that even Hansen's scenario remains plausible.

Until sound physics-based models are available to estimate ice sheet contributions to future sea level rise, past sea level rise may be our best guide to the future. During warming at the end of the last ice age sea level rise was dominated by the retreat of land-based ice sheets and occurred at an average rate of 1 to 2 meters (3.3 to 6.6 feet) per century for several thousand years.[26] There is no question, therefore, that large ice sheets can contribute to sea level rise at much higher rates than those projected by the IPCC; the question is rather a matter of timing. Traditionally, long lag times have been assumed for ice sheet response to warming, but this assumption is now receiving greater scrutiny. James Hansen and colleagues argue that the long lag times of sea level rise during past climates do not apply to the rapid change in atmospheric and oceanic temperatures today and that we should expect a much faster response from the polar ice sheets during this century.[27]

The warmest point of the last interglacial period, around 125,000 years ago, was about 1 to 2°C (1.8 to 3.6°F) warmer than the present global temperature for only a few centuries, yet saw sea levels 4 to 6 meters (13 to 20 feet) higher than at present.[28] Like the case today, large ice sheets occurred only on Greenland and Antarctica, so this period, known as the Eemian, is particularly relevant to our present conditions. Very recent research indicates that sea level rose at an average rate of 1.6 meters (5 feet) per century during this slightly warmer period.[29] Given that temperatures within the next century could be even warmer and lag times shorter, it is conceivable that even higher rates of sea level rise could occur within the current century.

Sea Level Rise Values for the Three Scenarios of Climate Change

Based on the expert opinion he has solicited and the writings of additional experts, this author judges that 2 meters (6.6 feet) of sea level rise is plausible during the twenty-first century under a scenario of rapid warming and ice sheet–dominated sea level rise. Although the choice of a particular number remains largely arbitrary—a sentiment expressed by several of the experts interviewed for this project—2 meters (6.6 feet) is physically plausible. This amount of sea level rise also corresponds conveniently to mapping programs that show coastline inundation in increments of 1 meter (3.3 feet), thus simplifying impact assessment (for example, see figure 3-10 accompanying box 3-3). Furthermore, 2 meters (6.6 feet) is not greatly more than a doubling of the upper bound of the 2001 IPCC sea level rise projection of 0.88 meter

(2.9 feet) for 2100.[30] In scenario 3, therefore, we adopt 2 meters (6.6 feet) for projected sea level rise at the end of the twenty-first century relative to 1990.

To obtain a projection of sea level rise for scenario 2, we use the projection of the 2001 IPCC report as a scaling function.[31] The upper end of the projection is 0.23 meter (0.75 foot) in 2040 and 0.88 meter (2.9 feet) in 2100; the ratio of 0.75 to 2.9 (0.75 ÷ 2.9) is 0.26. Multiplying this ratio by the posited rise for scenario 3 for the end of the twenty-first century (0.26 × 6.6) yields a sea level rise projection of 0.52 meter (1.7 feet) for the year 2040 relative to 1990 in scenario 2.

As stated previously, these sea level rise scenarios are not predictions and should not be taken as such nor used in ways other than are consistent with the purpose and intent of this volume. It is also important to keep in mind that regardless of how high the sea rises by the end of this century, many more centuries will pass before sea level reaches a balance with the change in temperature. Sustained warming of about 3°C (5.4°F) would eventually eliminate the Greenland Ice Sheet in future centuries, ultimately raising sea level by about 6 meters (20 feet); contributions from Antarctica would increase the total even more.[32]

Summaries of the Three Scenarios

This section provides brief summaries of the three climate scenarios. More details on regional changes and the impacts of sea level rise follow. The contribution of IPCC Working Group II to the *AR4* describes future global impacts for different systems resulting from progressively higher warming.[33] Tables 3-2 and 3-3 summarize these findings in aggregate, with markers for scenarios 1, 2, and 3.

Climate Scenario 1: Expected Climate Change

The average change obtained in IPCC projections based on the SRES A1B emissions scenario is realized without abrupt changes or other great surprises. By 2040 average global temperature rises 1.3°C (2.3°F) above the 1990 average. Warming is greater over land masses and increases from low to high latitudes. Generally, the most damaging local impacts occur at low latitudes because of ecosystem sensitivity to altered climate and high human vulnerability in developing countries, and in the Arctic because of particularly large temperature changes at high northern latitudes. Global mean sea level increases by 0.23 meter (0.75 foot), causing damage to the most vulnerable coastal wetlands with associated negative impacts on local fisheries,

Table 3-2. Examples of Global Impacts Projected for Increasing Change in Global Average Surface Temperature, Assuming No Mitigation or Adaptation Measures for Climate Change

	Global Mean Annual Temperature Change Relative to 1990 (°C)												
	0.0	0.5	1.0	1.5	2.0	2.5	3.0	3.5	4.0	4.5	5.0	5.5	6.0
				SC1 +1.3° C +23 cm SLR		SC2 +2.6° C +43 cm SLR						SC3 +5.6° C +2 m SLR	
Water	General increased irrigation demand for water →												
	Increased water availability in moist tropics & high latitudes												
								Up to ⅕ of world population affected by increased flood events →					
	Decreased water availability & increased drought in mid-latitudes & semi-arid low latitudes →												
	0.4 to 1.7 bn people with increased water scarcity →				1.0 to 2.0 bn people with increased water scarcity →					1.1 to 3.2 bn people with increased water scarcity →			
Food		Crop yield potential starts to decrease at low latitudes →						Adaptive capacity exceeded for all crops at low latitudes					
		Crop yield potential starts to increase at mid/high latitudes →						Yields of all crops decline at mid/high latitudes →					
		-10 to +30 mn more people at risk of hunger →			-10 to +10 mn more people at risk of hunger →				-30 to + 120 mn more people at risk of hunger →				

Coasts	Increased damage from floods and storms →		
		~30% loss of coastal wetlands →	
	0 to 3 mn more people at risk of flooding →	1 to 15 mn more people at risk of flooding →	
Health	Changed distribution of some infectious disease vectors & allergenic pollen species →		
		Increased burden from malnutrition, diarrheal, cardio-respiratory & infectious diseases	
		Increased morbidity & mortality from heat waves, floods, & droughts →	
		Substantial burden on health services →	
Abrupt events	Local retreat of ice in Greenland and West Antarctica →	Long-term commitment to several m SLR due to ice sheet loss →	Long-term commitment to SLR implying reconfiguration of coastlines worldwide and inundation of low-lying areas
		Changes in marine and terrestrial ecosystems due to MOC weakening →	Further weakening of MOC, possibly causing net cooling in northern high latitudes near Greenland and northwestern Europe

Source: Adapted from table SPM-1 in IPCC, "Summary for Policymakers," in *Climate Change 2007: Impacts, Adaptation, and Vulnerability* (see note 1 for complete bibliographical information), with additional information drawn from other chapters of the full report.

Abbreviations: dec = decrease, inc = increase, mn = million, bn = billion, m = meters, cm = centimeters, MOC = Meridional Overturning Circulation, SLR = sea level rise, WAIS = West Antarctica Ice Sheet.

Notes: Placement of the beginning of an impact indicates at what temperature the impact begins. Arrows indicate increasing levels of impact. Most socioeconomic impacts are time dependent (they assume certain levels of population, income, and technology, and these vary with time). Entries for these impacts in the table are for global average temperature levels projected under the SRES A2 scenario. Estimates for numbers of people with increased water scarcity, hunger, and coastal flooding cover the range of SRES scenarios A1FI, A2, B1, and B2. Estimations of numbers of people affected by water scarcity and sea level rise are relative to a future without climate change.

Impacts associated with sea level rise are adjusted to correspond with the three scenarios outlined in this chapter. This is, therefore, not an exact representation of the findings of the *AR4* and should not be construed as such.

Table 3-3. Examples of Regional Impacts Projected for Increasing Change in Global Average Surface Temperature

	Global Mean Annual Temperature Change Relative to 1990 (°C)												
	0.0	0.5	1.0	1.5	2.0	2.5	3.0	3.5	4.0	4.5	5.0	5.5	6.0
				SC1 +1.3° C +23 cm SLR		SC2 +2.6° C +43 cm SLR						SC3 +5.6° C +2 m SLR	
Africa		75 to 250 mn people w/ inc water scarcity →			350 to 600 mn people w/ inc water scarcity - - - →								
Asia		0.1 to 1.2 bn people w/ inc water scarcity →		0.2 to 1.0 bn people w/ inc water scarcity - - - →									
				0 to 2 mn more people at risk of coastal flooding annually →			0 to 7 mn more people at risk of coastal flooding annually - - - →						
Australia/ New Zealand			Annual bleaching of Great Barrier Reef		3,000 to 5,000 more heat-related deaths a year			1-in-50 flood of Butler River Westport, NZ, increased by 40%					
		Decreased water security in S&E Aus. & E. NZ →		10–25% decrease Murray-Darling Basin River flow →					15–50% decrease Murray-Darling Basin River flow - - - →				
Europe		5–15% increase in water availability in north →						10–20% increase - - - →					
		0–25% decreased availability →					5–35% decrease - - - →						
		2–10% inc in wheat yield in north →			10–25% inc →		10–30% increase - - - →						
		3–4% inc in wheat yield in south →			-10% to +20% change in yield →				15% to +30% change - - - →				

Region	Impacts
Latin America	Inter-tropical glaciers disappear; Potential extinction of ~25% central Brazilian savanna tree species ⟶ Potential extinction of ~45% Amazonian tree species ⇢
	10 to 80 mn people w/ inc water scarcity ⟶ 80 to 180 mn people w/ inc water scarcity ⇢
North America	5–20% increase forest growth; 10–20% increase crop growth; ~5% increase ozone-related deaths; 3–8 times increase heat wave days in some cities
	Decreased space heating needs ⇢
	Increased space cooling needs ⇢
Polar regions	10–15% inc depth of seasonal thaw-Arctic permafrost ⟶ 15–25% increase ⟶ 30–50% increase; 10–50% Arctic tundra replaced by forest; 10–25% polar desert replaced by tundra
	20–35% reduction of Arctic permafrost area; 20–35% decrease in Arctic sea ice
Small islands	Alien species colonize mid/high lat islands; Agriculture losses up to 5% in high terrain islands & up to 20% in low terrain islands ⇢
	Coastal inundation & damage to infrastructure due to SLR ⇢

Source: IPCC, "Summary for Policymakers" in *Climate Change 2007: Impacts, Adaptation, and Vulnerability* (see note 1 of the text for complete bibliographical information).
Abbreviations: dec = decrease, inc = increase, mn = million, bn = billion, m = meters, cm = centimeters, MOC = Meridional Overturning Circulation, SLR = sea level rise.

seawater intrusion into groundwater supplies in low-lying coastal areas and small islands, and elevated storm surge and tsunami heights, damaging unprotected coastlines. Many of the affected areas have large, vulnerable populations requiring international assistance to cope with or escape the effects of sea level rise. Marine fisheries and agricultural zones shift poleward in response to warming, in some cases moving across international boundaries. The North Atlantic MOC is not affected significantly.

Regionally, the most significant climate impacts occur in the southwestern United States, Central America, sub-Saharan Africa, the Mediterranean region, the mega-deltas of South and East Asia, the tropical Andes, and small tropical islands of the Pacific and Indian oceans. The largest and most widespread impacts relate to reductions in water availability and increases in the intensity and frequency of extreme weather events. The Mediterranean region, sub-Saharan Africa, northern Mexico, and the southwestern United States experience more frequent and longer-lasting drought and associated extreme heat events, in addition to forest loss from increased insect damage and wildfires.

Overall, northern mid-latitudes see a mix of benefits and damages. Benefits include reduced cost of winter heating, decreased mortality and injury from cold exposure, and increased agricultural and forest productivity in wetter regions because of longer growing seasons, CO_2 fertilization, and fewer freezes. Negative consequences include higher cost of summer cooling, more heavy rainfall events, more heat-related death and illness, and more intense storms with associated flooding, wind damage, and loss of life, property, and infrastructure.

Climate Scenario 2: Severe Climate Change

Average global surface temperature rises at an unexpectedly rapid rate to 2.6°C (4.7°F) above 1990 levels by 2040, with larger warming over land masses and at high latitudes. Dynamical changes in polar ice sheets—changes in the rate of ice flow into the sea—accelerate rapidly, resulting in 0.52 meter (1.7 feet) of global mean sea level rise. Based on these observations and an improved understanding of ice sheet dynamics, climate scientists by this time express high confidence that the Greenland and West Antarctic ice sheets have become unstable and that 4 to 6 meters (13 to 20 feet) of sea level rise are now inevitable over the next few centuries. Water availability decreases strongly in the most affected regions at lower latitudes (dry tropics and subtropics), affecting about 2 billion people worldwide. The North Atlantic MOC slows significantly, with consequences for marine ecosystem productivity and fisheries. Crop yields decline significantly in the fertile river deltas because of sea level rise and damage from increased storm surges. Agriculture becomes non-viable in the dry subtropics, where irrigation becomes exceptionally difficult because of low water availability and increased soil salinization resulting from more rapid evaporation of water from irrigated fields. Arid regions at low latitudes expand, taking previously marginally productive croplands out of production. North Atlantic fisheries are affected by significant slowing of the North Atlantic MOC. Globally, there is widespread coral bleaching, ocean acidification, substantial loss of coastal nursery wetlands, and warming and drying of tributaries that serve as breeding grounds for anadromous fish—ocean-dwelling fish that breed in freshwater, such as salmon. Because of a dramatic decrease in the extent of Arctic sea ice, the Arctic marine ecosystem is dramatically altered and the Arctic Ocean is navigable for much of the year.

Developing nations at lower latitudes are affected most severely because of climate sensitivity and low adaptive capacity. Industrialized nations to the north experience clear net harm and must divert greater proportions of their wealth to adapting to climate change at home.

Climate Scenario 3: Catastrophic Climate Change

Between 2040 and 2100 the impacts associated with climate scenario 2 progress and large-scale singular events of abrupt climate change occur. The average global temperature rises to 5.6°C (10.1°F) above 1990 levels, with greater warming over land masses and at higher latitudes. Because of continued acceleration of dynamical polar ice sheet changes, global mean sea level rises 2 meters (6.6 feet) relative to 1990, rendering low-lying coastal regions uninhabitable, including many large coastal cities. The large fertile deltas of the world become largely uncultivable because of inundation and more frequent and higher storm surges that reach farther inland. The North Atlantic MOC collapses at mid-century, generating large-scale disruption of North Atlantic marine ecosystems and associated fisheries. Northwestern Europe experiences colder winters, shorter growing seasons, and lower crop yields than those of the twentieth century.

Outside northwestern Europe and the northern North Atlantic Ocean, the MOC collapse increases average temperatures in most regions and reorganizes precipitation patterns in unpredictable ways, hampering water resource planning around the world and drying out existing grain-exporting regions. Southern Europe and the Mediterranean region remain warmer than the twentieth-century average and continue to experience hotter, drier summers with more frequent and longer heat waves, more frequent and larger wildfires, and lower crop yields. Agriculture in the traditional breadbaskets is severely compromised by alternating persistent drought and extreme storm events that bring irregular severe flooding. Crops are physiologically stressed by temperatures and grow more slowly, even when conditions are otherwise favorable. Even in many regions with increased precipitation, summertime soil moisture is reduced by increased evaporation. Breadbasket-like climates shift strongly northward into subarctic regions with traditionally small human populations and little infrastructure, including roads and utilities. Furthermore, extreme year-to-year climate variability in these regions makes sustainable agriculture difficult on the scale needed to feed the world population.

Mountain glaciers are virtually gone and annual snowpack dramatically reduced in regions where large human populations have traditionally relied on glaciers and annual snowfall for water supply and storage, including Central Asia, the Andes, Europe, and western North America. Arid regions expand rapidly, overtaking regions that traditionally received sufficient annual rainfall to support dense populations. The dry subtropics—including the Mediterranean region, much of Central Asia, northern Mexico, much of South America, and the southwestern United States—are no longer inhabitable. Not only is the area requiring remote water sources for habitability dramatically larger than in 1990, but such remote sources are much less available because mountain glaciers and snowlines have retreated dramatically. Half of the world's human population experiences persistent water scarcity.

Locally devastating weather events are the norm for coastal and mid-latitude continental locations, where tropical and mid-latitude storm activity and associated wind and flood damage becomes much more intense and occurs annually, leading to frequent losses of life, property, and infrastructure in many countries every year. Whereas water availability and loss of food security disproportionately affect poor countries at lower latitudes, extreme weather events are more or less evenly distributed, with perhaps greater frequency at mid-latitudes because of stronger extratropical storm systems, including severe winter storms.

General Patterns of Projected Climate Change

This section reviews general patterns of climate change as projected by the IPCC *Fourth Assessment Report* (*AR4*). The purpose is to provide a general template of regional patterns of climate impacts at subcontinental scales, over which to lay the generalities just described for the three scenarios. Unless otherwise indicated, the results described in this section are extracted from chapters 10 and 11, *Contribution of Working Group I to the Fourth Assessment Report,* which present projections of future climate change based on modeling experiments using mostly aggregated results of up to twenty-one different global climate models.[34] Changes are presented as averages of all the models used in an analysis.

Temperature

All models in the *AR4* show global surface warming in proportion to the amount of man-made greenhouse gases released to the atmosphere. For the SRES A1B emissions scenario, average global surface warming relative to 1990 is about 1.3°C (2.3°F) in 2040 and 2.8°C (5.0°F) in 2100. It is essential to put these global averages into geographic context, as changes are far from uniform globally. Temperature over land, particularly in continental interiors, warms about twice as much as the global average, as surface temperatures rise more slowly over the oceans. High northern latitudes also warm about twice as fast as the global average. Moreover, the average change in any given location is not a smooth increase over time. Rather, it is associated with larger extremes, leading to generally fewer freezes, higher incidence of hot days and nights, and more heat-related impacts, such as heat waves, droughts, and wildfires. Larger warming at high northern latitudes leads to faster thawing of permafrost, with consequent infrastructure damage such as collapsed roads and buildings, coastal erosion and feedbacks that amplify climate change such as methane and CO_2 release from thawed organic soils.[35] There are also seasonal differences, with winter temperatures rising more rapidly than summer temperatures, especially at higher latitudes. Wintertime warming in the Arctic over the twenty-first century is projected to be three to four times greater than the global wintertime average warming, resulting in much faster loss of ice cover and associated impacts such as faster sea level rise. (More regional detail is provided in box 3-2.)

Precipitation

Under the A1B emissions scenario, global average precipitation increases by 2 percent in 2040 and 5.5 percent in 2090 relative to 1990 levels. Because some regions experience substantially decreased precipitation, a global change of a few percentage points translates into changes greater than 20 percent for particular areas. Both extreme drought and extreme rainfall events are expected to become more frequent as a result of this intensification

Box 3-2. IPCC Findings for Regional Climate Projections

The following excerpts from the Executive Summary of chapter 11 of the *Contribution of Working Group I to the Fourth Assessment Report* summarize robust findings on projected regional changes over the twenty-first century.[1] The IPCC deems these changes likely (greater than 66 percent likelihood) to very likely (greater than 90 percent likelihood), taking into account the uncertainties in climate sensitivity and SRES emissions trajectories of the SRES B1, A1B, and B2 emissions scenarios.

Africa

Warming is very likely to be larger than the global annual mean warming throughout the continent and in all seasons, with drier subtropical regions warming more than the moister tropics. Annual rainfall is likely to decrease in much of Mediterranean Africa and the northern Sahara, with a greater likelihood of decreasing rainfall as the Mediterranean coast is approached. Rainfall in southern Africa is likely to decrease in much of the winter rainfall region and western margins. There is likely to be an increase in annual mean rainfall in East Africa. It is unclear how rainfall in the Sahel, the Guinean Coast, and the southern Sahara will evolve.

Mediterranean and Europe

Annual mean temperatures in Europe are likely to increase more than the global mean. Seasonally, the largest warming is likely to be in northern Europe in winter and in the Mediterranean area in summer. Minimum winter temperatures are likely to increase more than the average in northern Europe. Maximum summer temperatures are likely to increase more than the average in southern and central Europe. Annual precipitation is very likely to increase in most of northern Europe and decrease in most of the Mediterranean area. In central Europe, precipitation is likely to increase in winter but decrease in summer. Extremes of daily precipitation are very likely to increase in northern Europe. The annual number of precipitation days is very likely to decrease in the Mediterranean area. Risk of summer drought is likely to increase in central Europe and in the Mediterranean area. The duration of the snow season is very likely to shorten, and snow depth is likely to decrease in most of Europe.

Asia

Warming is likely to be well above the global mean in Central Asia, the Tibetan Plateau and northern Asia, above the global mean in East Asia and

Box 3-2. IPCC Findings for Regional Climate Projections (*continued*)

South Asia, and similar to the global mean in Southeast Asia. Precipitation in boreal winter is very likely to increase in northern Asia and the Tibetan Plateau, and likely to increase in eastern Asia and the southern parts of Southeast Asia. Precipitation in summer is likely to increase in northern Asia, East Asia, South Asia, and most of Southeast Asia, but is likely to decrease in Central Asia. It is very likely that heat waves/hot spells in summer will be longer, more intense, and more frequent in East Asia. Fewer very cold days are very likely in East Asia and South Asia. There is very likely to be an increase in the frequency of intense precipitation events in parts of South Asia and in East Asia. Extreme rainfall and winds associated with tropical cyclones are likely to increase in East Asia, Southeast Asia, and South Asia.

North America

The annual mean warming is likely to exceed the global mean warming in most areas. Seasonally, warming is likely to be largest in winter in northern regions and in summer in the Southwest. Minimum winter temperatures are likely to increase more than the average in northern North America. Maximum summer temperatures are likely to increase more than the average in the Southwest. Annual mean precipitation is very likely to increase in Canada and the northeast United States, and likely to decrease in the Southwest. In southern Canada, precipitation is likely to increase in winter and spring but decrease in summer. Snow season length and snow depth are very likely to decrease in most of North America except in the northernmost part of Canada, where maximum snow depth is likely to increase.

Central and South America

The annual mean warming is likely to be similar to the global mean warming in southern South America but larger than the global mean warming on the rest of the continent. Annual precipitation is likely to decrease in most of Central America and in the southern Andes, although changes in atmospheric circulation may induce large local variability in precipitation response in mountainous areas. Winter precipitation in Tierra del Fuego and summer precipitation in southeastern South America is likely to increase. It is uncertain how annual and seasonal mean rainfall will change over northern South America, including the Amazon forest. However, there is qualitative consistency among the simulations in some areas (rainfall increasing in

(continued)

Box 3-2. IPCC Findings for Regional Climate Projections (*continued*)

Ecuador and northern Peru and decreasing at the northern tip of the continent and in southern northeast Brazil).

Australia and New Zealand

Warming is likely to be larger than that of the surrounding oceans, but comparable to the global mean. The warming is less in the south, especially in winter, with the warming in the South Island of New Zealand likely to remain less than the global mean. Precipitation is likely to decrease in southern Australia in winter and spring. Precipitation is very likely to decrease in southwestern Australia in winter. Precipitation is likely to increase in the west of the South Island of New Zealand. Changes in rainfall in northern and central Australia are uncertain. Increased mean wind speed is likely across the South Island of New Zealand, particularly in winter. Increased frequency of extreme high daily temperatures in Australia and New Zealand, and a decrease in the frequency of cold extremes is very likely. Extremes of daily precipitation are very likely to increase, except possibly in areas of significant decrease in mean rainfall (southern Australia in winter and spring). Increased risk of drought in southern areas of Australia is likely.

Polar Regions

The Arctic is very likely to warm during this century more than the global mean. Warming is projected to be largest in winter and smallest in summer. Annual arctic precipitation is very likely to increase. It is very likely that the relative precipitation increase will be largest in winter and smallest in summer. Arctic sea ice is very likely to decrease in its extent and thickness. It is uncertain how the Arctic Ocean circulation will change. The Antarctic is likely to warm and the precipitation is likely to increase over the continent. It is uncertain to what extent the frequency of extreme temperature and precipitation events will change in the polar regions.

Small Islands

Sea levels are likely to rise on average during the century around the small islands of the Caribbean Sea, Indian Ocean, and northern and southern Pacific Ocean. The rise will likely not be geographically uniform, but large deviations among models make regional estimates across the Caribbean, Indian, and Pacific oceans uncertain. All Caribbean, Indian Ocean, and North

Box 3-2. IPCC Findings for Regional Climate Projections (*continued*)

and South Pacific islands are very likely to warm during this century. The warming is likely to be somewhat smaller than the global annual mean. Summer rainfall in the Caribbean is likely to decrease in the vicinity of the Greater Antilles, but changes elsewhere and in winter are uncertain. Annual rainfall is likely to increase in the northern Indian Ocean with increases likely in the vicinity of the Seychelles in December, January, and February, and in the vicinity of the Maldives in June, July, and August, while decreases are likely in the vicinity of Mauritius in June, July, and August. Annual rainfall is likely to increase in the equatorial Pacific, while decreases are projected by most models for just east of French Polynesia in December, January, and February.

1. J. H. Christensen and others, "Regional Climate Projections," in *Climate Change 2007: The Physical Science Basis,* edited by Susan Solomon and others (Cambridge University Press, 2007), pp. 847–940.

of the global water cycle. Increased precipitation generally prevails in the tropics and at high latitudes, particularly over the tropical Pacific and Indian oceans during the Northern Hemisphere winter and over South and Southeast Asia during the Northern Hemisphere summer. Decreased precipitation prevails in the subtropics and mid-latitudes, with particularly strong decreases in southwestern North America and Central America, southern South America (parts of Chile and Argentina), southern Europe and the Mediterranean region in general (including parts of the Middle East), and in northern and southern Africa. Central America experiences the largest decline in summer precipitation. The main areas projected to experience greater drought are the Mediterranean region, Central America, Australia and New Zealand, and southwestern North America.[36]

Decreases in precipitation and related water resources are projected to affect several important rain-fed agricultural regions, particularly in South and East Asia, in Australia, and in northern Europe. Although monsoon rainfall is projected to increase in South and Southeast Asia, this extra rain may not provide benefits, as rain is already plentiful at this time of year. However, the added rainfall will likely increase damage from flooding. Notably, a decrease in summer precipitation is projected for Amazonia, where the world's largest complex of wet tropical forest depends on high year-round precipitation.[37]

Two important correlates of precipitation are annual runoff (surface water flow) and soil moisture. These parameters are critical to water supply for consumption and irrigation and to the ability of soil to support crop production and natural ecosystems. Soil moisture generally corresponds with precipitation, but declines in some areas where precipitation increases because warmer temperatures lead to greater evaporation. The biggest changes in soil moisture include a strong increase in a narrow band of equatorial Africa and a moderate increase in a band extending from northern and eastern Europe into Central Asia. Soil drying is more widespread; soil moisture decreases by 10 percent or greater over much of the United States, Mexico and Central America, southern Europe and the Mediterranean basin in general (including parts of the Middle East), southern Africa, the Tibetan Plateau, and across much of northern Asia.

Runoff follows a pattern very similar to precipitation, with increases in high northern latitudes and parts of the tropics, including Central, South, and Southeast Asia, tropical eastern Africa, the northern Andes and the east-central region of South America around Uruguay, and extreme southern Brazil (see figure 3-3, color plates). The strongest decreases occur in the southwestern United States, Central America, the Mediterranean region (including southern Europe, northern Africa, and the Middle East), southern Africa, and northeastern South America, including Amazonia.

Regional Sensitivity to Climate Change

A given change in climate such as a degree of warming or a 10 percent change in precipitation does not affect all regions the same way. It may be useful, therefore, to examine how sensitive different regions might be to changes in temperature or precipitation. From a security perspective it would then be useful to compare regional sensitivity to the distribution of global population density and to regions that are important for crop production.

The global distribution of population shown in figure 3-4 has a striking correspondence to the distribution of land that is currently suitable for producing rain-fed crops.[38] This pattern holds for the United States, even though extensive irrigation augments precipitation to increase crop yields, implying that historical rainfall patterns remain the primary determinants of regional agricultural productivity and population density.

Some regions experience a very stable climate, and natural and human systems have developed around this stability; in such regions even a small change may generate significant impacts. For instance, in wet tropical systems moderate decreases in precipitation may lead to the collapse of productive rainforests.[39] Alternatively, settlements and infrastructure in wet tropical regions may be damaged by increased flooding from small increases in precipitation during the rainy season. Semi-arid regions that are already marginal for supporting natural and human systems may be rendered uninhabitable by small decreases in precipitation or runoff. In contrast, regions with historically large climate variability require larger changes of future climate to move natural and human systems beyond the bounds of the climate extremes to which they have adapted. For instance, in spite of great natural climate variability, the Arctic is expected to be heavily impacted by climate change because the degree of warming is projected to be large.

A climate change index describing regional sensitivity to climate change has been described by M. B. Baettig, M. Wild, and D. M. Imboden, shown in figure 3-5.[40] The areas most sensitive to a combination of projected temperature and precipitation change relative to natural historical variability are in tropical Central and South America, tropical and southern Africa, Southeast Asia, and the polar regions. The Mediterranean region, China, and the western United States show intermediate levels of sensitivity. Marginal agricultural lands—such as those in the southwestern United States, Central America, sub-Saharan Africa, southern Europe, Central Asia, including the Middle East, and eastern China—generally show intermediate to high climate sensitivity. Most of these regions also have large human populations. Also of note, the most affected region of South America completely covers the Amazonian rainforest, which is projected to become relatively drier. Reduced productivity of this forest would have strong feedbacks on global climate by releasing carbon to the atmosphere and would result in massive loss of biodiversity, including economically important species.[41]

Extreme Weather Events

In general, the IPCC projects an increased incidence of extreme weather events.[42] Droughts, flash floods, heat waves, and wildfires are all projected to become more frequent and more intense in regions where such events are already common. Intense tropical and mid-latitude storms with heavier precipitation and higher wind speeds are also projected. There is evidence that many of these events already occur more frequently and have become more intense.[43] Projections indicate fewer cold spells and a decrease in the frequency of low-intensity storms. As a consequence, the total number of storms decreases globally even as the number of intense storms increases.

Extreme precipitation and drought. In general, the IPCC projects that a larger fraction of total precipitation will fall during extreme events, especially in the moist tropics and in mid- and high latitudes, where increased mean precipitation is projected. Regionally, extremes are expected to increase more than the means. Even in areas projected to become drier, the average intensity of precipitation may increase because of longer dry spells and greater accumulation of atmospheric moisture between events. This portends increased incidence and duration of drought, punctuated by extreme precipitation, which may be either rainfall or snowfall, depending on latitude and season. In general, the risk of drought is expected to increase during summers in the continental interiors.

Some tropical and subtropical regions experience monsoons, distinct rainy seasons during which prevailing winds transport atmospheric moisture from the tropical oceans. The Asian, African, and Australian monsoons are projected to bring increased rainfall to certain regions of these continents. Because this rain falls during what is already the rainy season, it may cause more flooding without bringing additional benefits. In Mexico and Central America, the monsoon is projected to bring less precipitation to the region, contributing to the increased drought generally projected for the region.

Temperature extremes. Hotter temperature extremes and more frequent, more intense, and longer-lasting heat waves are robust projections of the models examined by the IPCC, portending increased heat-related illness and mortality. Growing seasons will also become longer because of earlier spring warming and later fall cooling, but crops will face greater heat stress and associated drought during the growing season. Cold spells will become less frequent, causing fewer deaths and economic losses associated with cold weather.

Tropical and mid-latitude storms. Projected patterns of change are similar for both tropical cyclones, including typhoons and hurricanes, and extratropical cyclones (mid-latitude storms). Tropical storms may become less frequent overall, yet are expected to reach higher peak wind speeds and bring greater precipitation on average. The decrease in frequency is likely to result from fewer weak tropical storms, whereas intense tropical storms may become more frequent with warming. Similarly, mid-latitude storms may become less frequent in most regions yet more intense, with more damaging winds and greater precipitation. Intensification of winter mid-latitude storms may bring more frequent severe snowstorms, such as those experienced in the north-central United States in February and March of 2007. Near coasts, both tropical and mid-latitude storms will increase wave heights and storm surge heights, increasing the incidence of severe coastal flooding (see section on abrupt sea level rise on page 80).

Regions affected by tropical storms, including typhoons and hurricanes, include all three coasts of the United States; all of Mexico and Central America; the Caribbean islands; East, Southeast, and South Asia; and many South Pacific and Indian Ocean islands. Although tropical storms are very rare in the South Atlantic, in 2004 Hurricane Catarina became the only hurricane to strike Brazil in recorded history.[44] Similarly, it is unusual for tropical storms to make landfall in Europe, yet in 2005 the remnants of Hurricane Vince became the first tropical storm on record to make landfall on the Iberian Peninsula.[45] In June 2007 Cyclone Gonu, the first category 5 hurricane documented in the Arabian Sea, temporarily halted shipping through the Strait of Hormuz, the primary artery for exporting Persian Gulf oil.[46] Whether such historical aberrations are related to global warming remains uncertain, but extreme weather events exceeding historical precedents should be expected as a general consequence of climate change. Hence, such events illustrate the consequences of climate change for a given region.

Singular, Abrupt Events

With the assumptions of scenario 3, the probability and consequences of abrupt events move beyond the bounds of the assumptions of the IPCC projections. This departure is necessary as the potential consequences of large-scale abrupt events are of particular concern, yet the science for projecting and assessing them remains significantly underdeveloped.[47] To assess the consequences of such events, therefore, we draw upon the author's own assessment of a few particularly informative but uncertain studies.

Collapse of the North Atlantic meridional overturning circulation. The Gulf Stream is part of the North Atlantic meridional overturning circulation (MOC; also known as the thermohaline circulation or the Atlantic conveyor belt). The Gulf Stream transports warm tropical surface water from the equatorial North Atlantic Ocean northward along the East Coast of North America and then eastward toward northern Europe. From here the water flows north toward southern Greenland and the North Sea. Throughout this journey, the surface water cools and consequently becomes denser, eventually causing it to sink in the far North Atlantic near Greenland and flow southward at depth; this sinking drives the overturning circulation and sustains the continued transport of heat from the equator northward. This ocean transport of heat may warm the climate of northwestern Europe by several degrees. Global warming is thought to present a risk of shutting down the MOC by warming and freshening northern North Atlantic surface water as a result of Arctic ice melt, increased Arctic river runoff, and increased precipitation over the North Atlantic. Because warm water is less dense than cold water and freshwater is less dense than saltwater, global warming could slow or stop the MOC by reducing the tendency for water to sink in the far North Atlantic.[48]

Collapse of the MOC has often been described as a low-probability, high-impact event. In fact, however, there is tremendous variation among models and in experts' judgments regarding the probability of such an event.[49] There is even some disagreement about whether the MOC actually warms Europe significantly.[50] Moreover, there has been little investigation of the potential consequences of such an event, and it remains unclear whether it would indeed be of great consequence.[51] It is therefore all the more important not to regard the scenario outlined here as a prediction. Our purpose is to explore the possibility that collapse of the MOC could have a large impact, as such an outcome is widely considered plausible, if improbable.[52]

According to the IPCC, models that accurately represent past and current climate project a slowing of the Atlantic MOC of up to 60 percent, but none indicates a complete shutdown during the twenty-first century. As a result, the IPCC places the likelihood of a shutdown of the MOC during the twenty-first century at not more than 10 percent.[53] In the IPCC models, slowing of the MOC of up to 60 percent does not produce a cooling of Europe, as the warming effect of increasing atmospheric greenhouse gases offsets the cooling effect of the slower MOC. If, however, the rate of warming and loss of polar ice has been underestimated by the models, as assumed in scenario 3, then the chance of a collapse of the MOC during this century could be considerably higher. Should an abrupt collapse occur, a cooling of the North Atlantic region, including northwestern Europe, is more likely.[54] We therefore consider the potential consequences of North Atlantic MOC collapse in scenario 3.

As it is not possible to estimate the timing of MOC collapse for a given degree of warming, we arbitrarily assume a collapse during the 2050s, with attendant impacts occurring in subsequent decades of the twenty-first century (and beyond). This approach mirrors that of Nigel W. Arnell, who sim-

Figure 3-3. Model-Projected Changes in Annual Water Runoff for 2050, Percentage Change Relative to the Average over 1900–70

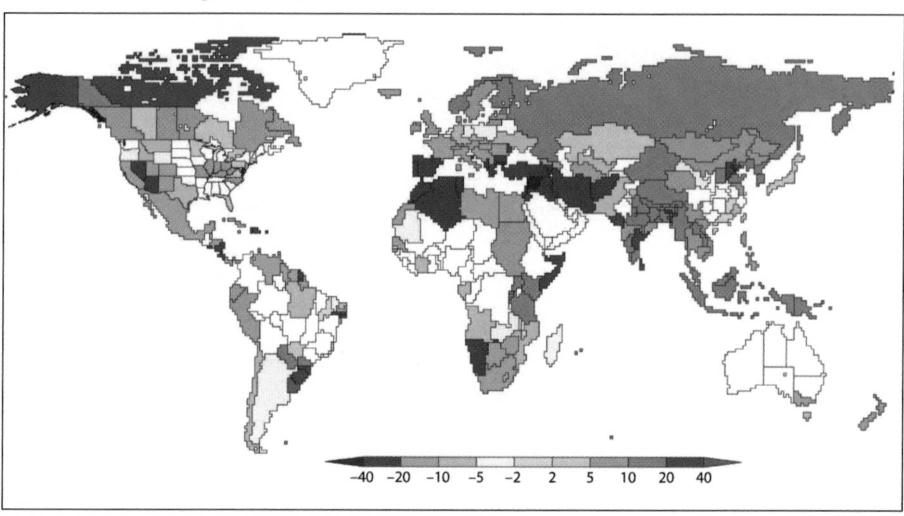

Source: Updated from P. C. D. Milly, K. A. Dunne, and A. V. Vecchia, "Global Pattern of Trends in Streamflow and Water Availability in a Changing Climate," *Nature* 438 (2005): 347–50.

a. Any color indicates that more than 66 percent of models agree on sign of change.

Figure 3-4. Distribution of Current Global Population Density

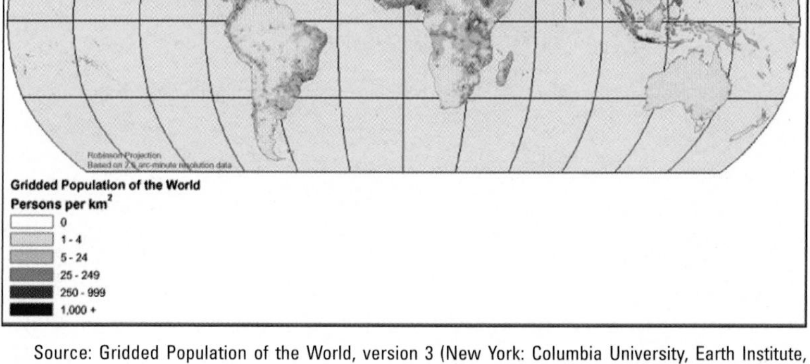

Source: Gridded Population of the World, version 3 (New York: Columbia University, Earth Institute, Center for International Earth Science Information Network (CIESIN)) (http://sedac.ciesin.columbia.edu/gpw/).

Figure 3-5. Index of Regional Sensitivity to Projected Changes in Temperature and Precipitation

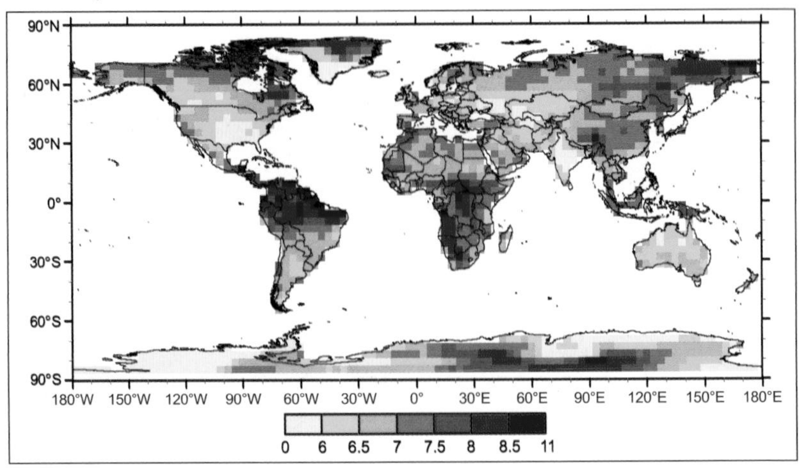

Source: The aggregated CCI in figure 2 of M. B. Baettig, M. Wild, and D. M. Imboden, "Climate Change Index: Where Climate Change May Be Most Prominent in the 21st Century," *Geophysical Research Letters* 34 (2007):L01705, doi:10.1029/2006GL028159.

Figure 3-6. San Francisco Bay and the Sacramento–San Joaquin River Delta, Inundated by One Meter of Future Sea Level Rise Shown in Red

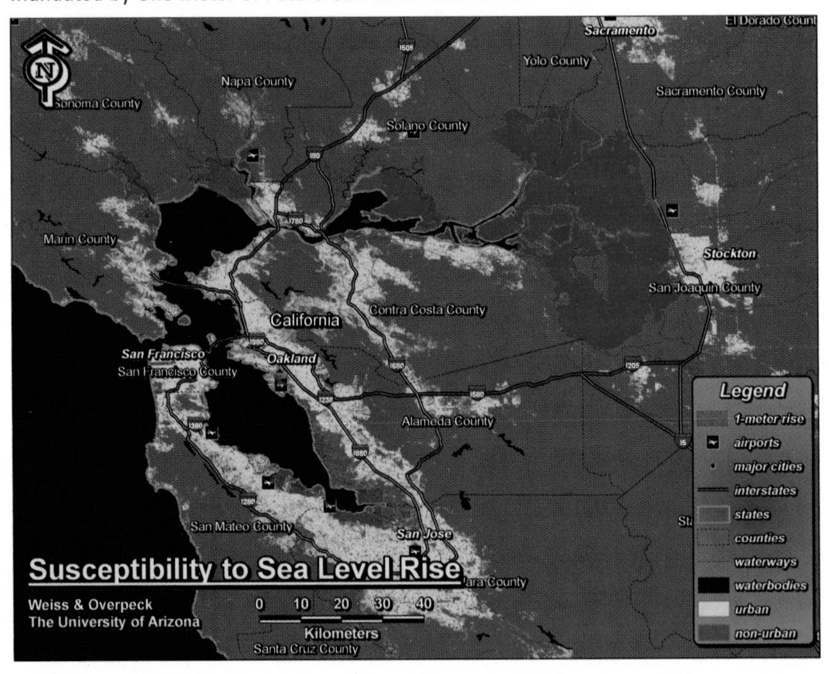

Source for figures 3-6, 3-7, 3-8, 3-9: J. L. Weiss and J. T. Overpeck, *Climate Change and Sea Level: Maps of Susceptible Areas.* Department of Geosciences Environmental Studies Research Laboratory, University of Arizona, 1997 (www.geo.arizona.edu/dgesl/research/other/climate_change_and_sea_level/sea_level_rise/sea_level_rise.htm).

Figure 3-7. Middle Atlantic Coast of the United States, Inundated by One Meter of Future Sea Level Rise Shown in Red

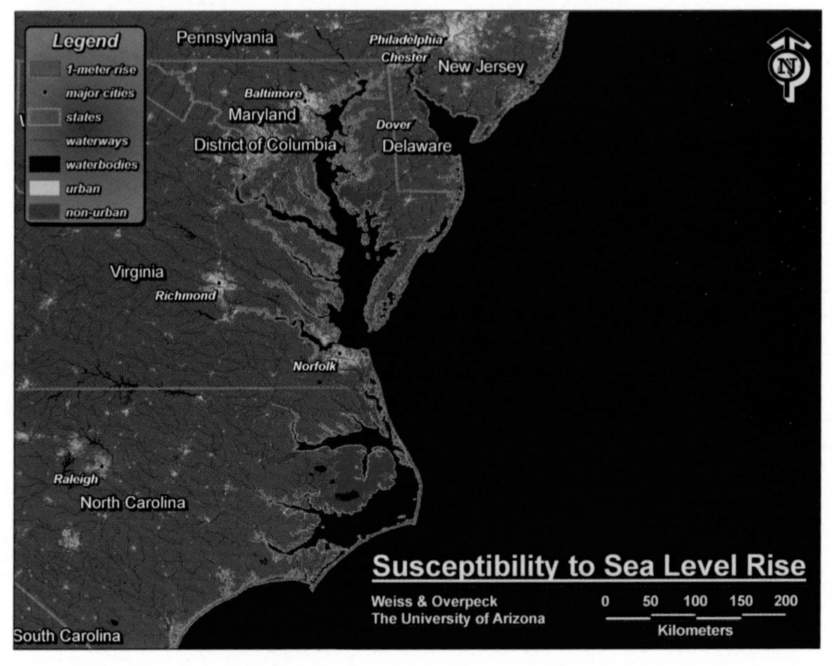

Figure 3-8. Southeastern Coastline of Australia, Inundated by One Meter of Future Sea Level Rise Shown in Red

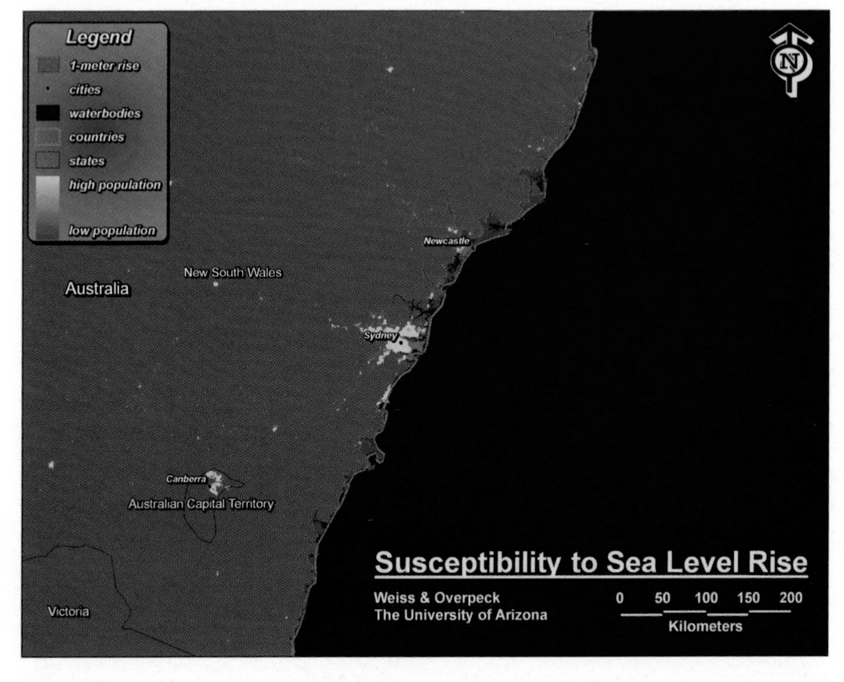

Figure 3-9. Mekong Delta, Vietnam, Inundated by One Meter (Bright Red) and Two Meters (Dark Red) of Future Sea Level Rise

Figure 3-10. Mekong Delta, Vietnam, with Population Density in the Low-Elevation Coastal Zone (LECZ)[a]

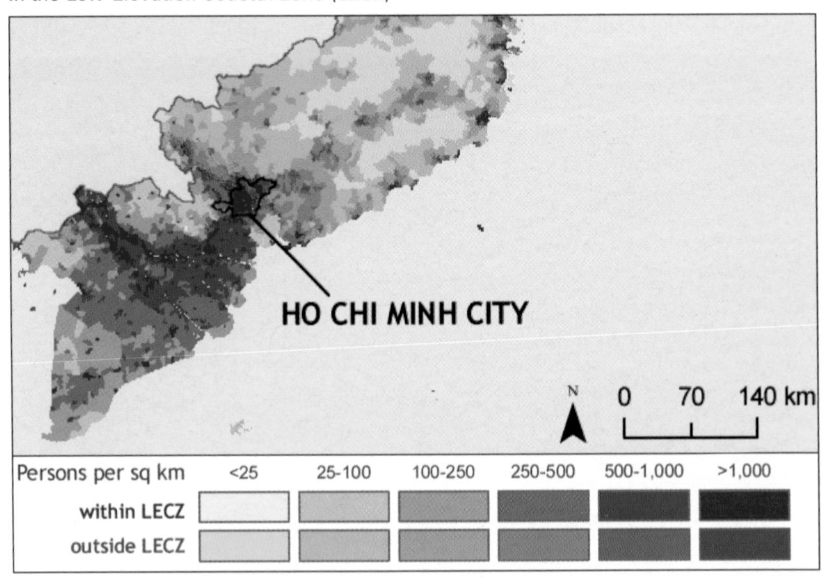

Source for figure 3-10: G. McGranahan, D. Balk, and B. Anderson, "The Rising Tide: Assessing the Risks of Climate Change and Human Settlements in Low Elevation Coastal Zones," *Environment & Urbanization* 19 (1997): 17–37 (adapted and used with permission based on Creative Commons 2.5 attribution. License: http://creativecommons.org/licenses/by/2.5).

a. The LECZ indicates an area below 10 meters elevation and shown in shades of red, with the population density outside the LECZ shown in shades of green.

ulated a shutdown of the North Atlantic MOC in a global circulation model in the year 2055 and followed its subsequent effects on water resources, energy use, human health, agriculture, and settlement and infrastructure.[55] Because there are few studies of this nature, we base the effects of an MOC collapse in scenario 3 on the results of that study. Arnell forced a global climate model with SRES scenario A2 and separately forced a shutdown of the MOC by imposing an artificial freshwater pulse in the North Atlantic.[56] Temperature change from the A2 scenario is similar to that of the A1B scenario until late in the twenty-first century, when A2 produces more warming than A1B. The impact of shutting down the MOC was compared to impacts of the A2 scenario without the freshwater pulse to shut down the MOC. It is important to understand that the MOC would not have shut down in the model if not for this artificially imposed freshwater pulse, an experimental manipulation applied solely to assess the potential impacts of an MOC collapse. If, however, the model underestimates the loss of polar ice as assumed in scenario 3, then it is reasonable to compensate for the underestimation by imposing additional freshwater input.

In general, MOC collapse resulted in cooler temperatures around the northern North Atlantic, with the largest effect centered south of Greenland and decreasing with distance from this central area. Areas of northwestern Europe cooled by as much as 3°C (5.4°F), with broader areas of Europe and northeastern North America cooling by 1 to 2°C (1.8 to 3.6°F). Many other parts of the world warmed because of a redistribution of heat from changes in ocean currents. Precipitation changes were more widespread than cooling, with attendant changes in runoff, drought, and flooding. The largest decreases in precipitation occurred in North Africa, the Middle East, Central America, the Caribbean, and northeast South America, including Amazonia. Intermediate decreases in precipitation were more widespread, including central North America, southern Greenland, central and southern Europe, central and southeast South America, Central and South Asia, western and southern Africa, and Australia. The largest increase in precipitation was centered on the southwestern United States, providing a net reduction in the number of people in the country under water stress. Increased precipitation also occurred in the eastern United States, Canada, East Africa, and northern, eastern, and Southeast Asia.

Several of the world's major grain-exporting regions, particularly in North America and South Asia, were affected by increased drought as a result of reduced precipitation after MOC collapse. In Europe, this trend would be exacerbated by lower temperatures and shorter growing seasons. Hence, global food markets would likely be affected by short supply and high prices. In Europe and northeastern North America, demand for heating fuel would increase because of colder winters. Although demand for cooling fuel would decrease in these regions, most other regions of the world would experience increased demand for cooling fuel. The cost of maintaining and adapting transportation infrastructure and demand for heating fuel would increase in northern Europe and northeastern North America, resulting in a southward shift of economic activity and population.[57]

Another consequence of a complete MOC collapse is likely to be an increase in sea level in the North Atlantic region, in addition to global mean sea level rise.[58] Model results and expert opinion suggest that this effect could add up to 1 meter (3.3 feet) of sea level rise in the Atlantic north of 45 degrees N, bringing total sea level rise for this region to 3 meters (10 feet) in our *catastrophic* scenario 3, with attendant coastal impacts (see discussion on abrupt sea level rise below).[59]

In general, the effects of accelerated global warming without MOC collapse are larger than the effects of MOC collapse. Broadly, however, accelerated climate change is expected to intensify current precipitation patterns, offering some degree of predictability and maintaining current geographic patterns of large-scale food production. By reorganizing precipitation patterns, MOC collapse may threaten major crop-producing regions with decreased precipitation, raising the possibility of major disruptions in global food supply.[60] It also appears to amplify the decrease of precipitation in Central America and Amazonia, threatening tropical forests and their dependent species with extinction and adding additional carbon to the atmosphere through large-scale forest dieback, amplifying the global greenhouse warming trend. Although water stress increases in parts of Africa and Asia, increased precipitation in East Africa and East and Southeast Asia results in a net of 1 billion fewer people under water stress with MOC collapse, but adds to flood hazards in these regions.

Abrupt sea level rise. The AR4 projects sea level rise in the range of 0.18 to 0.59 meter (0.6 to 1.9 feet) by the end of the century. As discussed earlier, however, this projection excludes an estimate of accelerated ice loss from the Greenland and Antarctic ice sheets (ice loss raises sea level even if the ice has not yet melted) and therefore cannot be considered either a best estimate or an upper bound for future sea level rise.[61] Moreover, the IPCC projections depict a gradual change in sea level over the next century, whereas abrupt and intermittent rises may be more likely, particularly for individual regions (see box 3-1). In the climate impacts scenarios outlined here, we assume that sea level rises 0.23 meter (0.75 foot) (scenario 1) or 0.52 meter (1.7 feet) (scenario 2) relative to 1990 by 2040, or 2 meters (6.6 feet) (scenario 3) relative to 1990 by 2100 (see table 3-1). As noted, under scenario 3, additional sea level rise of up to 1 meter (3.3 feet) would occur in the northern North Atlantic as a consequence of North Atlantic MOC collapse.[62] The possibility that extreme sea level rise could occur abruptly and unpredictably in this region (or others) should be considered in risk assessments.

Although it is safe to assume that greater sea level rise leads to relatively more severe impacts, studies of potential sea level rise impacts have not been conducted for most parts of the globe, and those that have been typically examine only one aspect of sea level impacts, such as beach erosion or storm surge height.[63] Sea level rise varies regionally, but future regional patterns are unpredictable at present.[64] Moreover, a lack of highly resolved global demographic data for coastal areas has hampered systematic assessment of coastal hazards.[65] Recent geographic population estimates indicate that about one-tenth of the world's population lives in coastal regions within 10 meters (33 feet) of sea level, and the global population continues to migrate coastward.[66] This estimate offers a general sense of the relative susceptibility of different regions to sea level rise impacts but cannot tell us how many people are likely to be directly impacted by sea level rise of the magnitude assumed in our scenarios (0.23 to 2.0 meters). In summary, it is currently extremely difficult to quantify future damage to humanity from sea level rise, although damage from a rise of 2 meters (6.6 feet) during the current century would clearly be catastrophic for many regions, including key areas within the United States.[67]

Sea level rise causes or contributes to several distinct types of impacts, including inundation, increased flooding from coastal storms, coastal erosion, saltwater intrusion into coastal water supplies, rising water tables, and coastal and upstream wetland loss with attendant impacts on fisheries and other ecosystem services.[68] Current distribution of natural and human coastal systems has been adapted to past extreme high tides and storm surges. Future sea level rise will inundate additional land not so adapted. Only the lowest-lying, unprotected areas will be extremely vulnerable to inundation within the time frame of our thirty-year scenarios. There are dozens of coastal cities worldwide in both industrialized and developing nations that lie at least partly below 1 to 2 meters (3.3 to 6.6 feet) elevation, but most of them have flood protection. Hence, inundation from extreme high tides alone might not rise to crisis proportions for most of these cities within the coming century, although enhanced defenses will be required to avoid increasing damages.

Inundation is a serious issue nonetheless for unprotected low-lying areas, including coastal wetlands that serve as natural nurseries for important fisheries, and productive agricultural lands situated on river deltas, a particularly sensitive problem for coastal aquifers and Asian mega-deltas.[69] Because of their inherently low elevations, proximity to the open sea, and general lack of flood protection, coastal wetlands are probably the most vulnerable of all natural systems to inundation and are also of underappreciated importance to society.[70] For example, about 75 percent of the commercial fish catch and 90 percent of recreational fish catch in the United States depends on wetlands that serve as nurseries and feeding grounds for fish and shellfish. Habitat loss and modification are the dominant causes of the worldwide decline in ocean fish catch during the past two decades.[71] One meter of sea level rise could eliminate or damage half of coastal wetlands globally, with the most vulnerable wetlands located along the Mediterranean and Baltic coasts and the Atlantic coasts of Central and North America, including the Gulf of Mexico.[72] Chronic saltwater inundation would devastate agricultural production as well, and the situation is similar for coastal groundwater supplies, which cannot be protected by levees or other surface-level devices.

Beyond the twenty-first century, sea level rise could far exceed 2 meters (6.6 feet), such that inundation eventually redraws coastlines altogether.[73] For the near term, however, more frequent and more severe flooding from coastal storms is likely to be the largest impact of sea level rise along low-lying coastlines.[74] Existing flood protection systems built to withstand extreme storm surges will be overcome much more frequently as local sea levels rise.[75] For example, levees around New Orleans were designed to withstand storm surges associated with category 3 hurricanes,[76] which historically attained heights of 2.8 to 3.7 meters (9.1 to 12.1 feet). Such defenses would be reduced effectively to category 2–level protection with 1 meter (3.3 feet) of sea level rise and category 1–level protection with 2 meters (6.6 feet) of sea level rise. Because weaker storms occur more frequently than the most intense storms, sea level rise portends a nonlinear increase in flood risk for protected areas in the absence of defense enhancement.[77] As another example, current flood defenses in New York City were designed to protect against a hundred-year flood—that is, the highest floodwaters expected to occur in a hundred-year period, on the basis of average past climate. However, 1 meter (3.3 feet) of sea level rise would lower the return interval of such a flood to as little as five years.[78] This estimate does not account for storm intensification, which would raise maximum storm surge and wave heights even further and is a predicted consequence of global warming.[79] The most critical areas of low-lying coastlines are cities and farmed deltas. Dozens of the world's most populous and culturally and economically important cities—New York, Miami, London, Copenhagen, Dublin, Sydney, Auckland, Shanghai, Bangkok, Calcutta, Dhaka, Alexandria, Casablanca, Lagos, Dakar, Dar es Salaam—are susceptible to sea level rise, as are some of the most important agricultural sites, such as the Sacramento, Ganges, Mekong, Yangtze, and Nile deltas.

Conclusion

Decisions as to whether to act and, if so, what actions to take to mitigate or prepare for the security implications of climate change will necessarily be made in the face of fundamental uncertainty about future climate conditions and their impacts. Uncertainty, however, is not tantamount to the inability to assess risk. Recent observations reveal a general tendency for climate projections to underestimate future change. As a result, current projections should be taken as the minimum change that society should expect to occur, and it should be recognized that there is genuine risk of much larger impacts than current projections imply. This volume embraces this view by developing three increasingly severe scenarios of future climate change impacts.

The least severe climate scenario accepts the best estimates of current IPCC projections for a mid-range greenhouse gas emissions scenario as the minimum amount of change to expect. The more severe scenarios assume that current projections significantly underestimate change because of the omitted positive feedback effects of warming and because of underestimated responsiveness of the climate system to warming. The most severe scenario, which plays out over a longer time period than the other two scenarios, additionally assumes that abrupt destabilization of polar land-based ice sheets leads to extreme sea level rise and the collapse of the North Atlantic meridional overturning circulation. The latter in turn leads to global-scale disruption of ecosystems and reorganizes the atmosphere in such a way that rainfall patterns shift away from major grain-exporting regions. These three scenarios provide the basis for assessing the national security implications of various plausible futures. In the following chapters, national security experts envision the possible consequence of these climate scenarios.

Box 3-3. Examples of Coastal Areas Susceptible to Sea Level Rise [see color plates for figures 3-6 – 3-10]

Many coastal cities and heavily populated mega-deltas around the world are at or near sea level and are therefore highly susceptible to impacts associated with sea level rise (see appendix B). Four examples are given here: the San Francisco Bay and Mid-Atlantic regions of the United States, the Australian cities of Sydney and Newcastle, and the Mekong Delta of Vietnam.

The United States has more major cities vulnerable to sea level rise than any other country (see appendix B). Significant portions of the U.S. Atlantic and Gulf coasts are vulnerable, as well as key sections of the Pacific Coast. The extreme vulnerability of the Mississippi River Delta recently gained notoriety in the aftermath of Hurricane Katrina. The Sacramento–San Joaquin River Delta is an inland delta connected to the ocean via San Francisco Bay (figure 3-6). It is one of the most agriculturally productive regions in the United States and has a population of about 400,000 residents. Most of the land in this delta lies below sea level and relies on more than 1,610 kilometers (about 1,000 miles) of levees to keep 280,000 hectares (700,000 acres) of land dry.[1] In 2004 a 108-meter (355-foot) section of an earthen levee on the Middle River collapsed, inundating 4,400 hectares (11,000 acres) of active cropland on a delta island called Jones Tract.[2] Repairing the levee and pumping the island dry took more than six months. Total losses and repair costs approached $100 million, and this figure would have been even higher had there been damage to the railroad, water pipeline, or petroleum pipeline that traverse the island. There are more than fifty other islands in the delta, most of which are also farmed and lie below sea level.[3]

The Atlantic Coast from Cape Cod southward is largely susceptible to sea level rise, especially the coastlines of New Jersey, Delaware, Maryland, Virginia, and North Carolina (figure 3-7). The coastlines of these states support strong fishing and tourism industries that are dependent on erosion-prone beaches and inundation-prone wetlands. Economically and ecologically important estuaries with extensive wetlands would be heavily impacted by 1 meter (3.3 feet) of sea level rise, including the Delaware Bay, the Chesapeake Bay–Potomac River, and Albemarle and Pamlico sounds. The Chesapeake Bay fishery is supported by coastal wetlands that would become inundated with only moderate sea level rise.[4] Major population centers in this region that are also vulnerable to inundation and increased storm surge include Baltimore, Washington, D.C., and the Virginia Beach–Norfolk–Newport News metropolitan area (figure 3-7).

Box 3-3. Examples of Coastal Areas Susceptible to Sea Level Rise (*continued*)

Australia is largely a coastal nation, with about 50 percent and 30 percent of its population located within 7 and 2 kilometers (4.3 and 1.2 miles), respectively, of the shoreline, and about 6 percent (1.25 million) of its residences in coastal zones below 5 meters (16.4 feet) elevation.[5] Sydney, Australia's most populous city, with 4.3 million residents, is situated partially below 1 meter (3.3 feet) elevation (figure 3-8). The nearby city of Newcastle, with about 0.5 million residents, lies mostly below 1 meter (3.3 feet) elevation. Other Australian cities susceptible to sea level rise include Melbourne, Adelaide, Cairns, and Darwin. Auckland, New Zealand, is also vulnerable (see appendix B).

Large river deltas, where 5 to 10 percent of the world's human population lives, are among the most vulnerable land masses to sea level rise.[6] The largest number of heavily populated and agriculturally productive mega-deltas are located in Asia. The Mekong River Delta in Vietnam, for example, has about 20 million inhabitants and supports a major portion of Asia's staple grain (rice) production (figures 3-9 and 3-10). Nearly the entire delta would be inundated by 1 meter (3.3 feet) of sea level rise; 2 meters (6.6 feet) would inundate significantly more land area and threaten Ho Chi Minh City, one of the most densely populated cities in Southeast Asia. Other major Asian cities situated on large deltas near sea level include Calcutta, Dhaka, Rangoon, Bangkok, Hanoi, and Shanghai. Similar circumstances prevail in mega-deltas on other continents as well, including the Nile, Niger, Amazon, and Mississippi deltas.

1. California Department of Water Resources, "Sacramento/San Joaquin Delta Atlas," 1995 (http://rubicon.water.ca.gov/delta_atlas.fdr/datp.html).

2. C. Souza, "Levee Break Losses Go into the Tens of Millions," *Ag Alert* (California Farm Bureau Federation), June 16, 2004.

3. California Department of Water Resources, "Sacramento/San Joaquin Delta Atlas."

4. S. M. Stedman and J. Hanson, "Habitat Connections: Wetlands, Fisheries and Economics," National Marine Fisheries Service (www.nmfs.noaa.gov/habitat/habitatconservation/publications/habitatconections/habitatconnections.htm).

5. K. Chen and J. McAneney, "High-Resolution Estimates of Australia's Coastal Population," *Geophysical Research Letters* 33 (2006): L16601, doi: 10.1029/2006GL026981.

6. J. P. Ericson and others, "Effective Sea-Level Rise and Deltas: Causes of Change and Human Dimension Implications," *Global and Planetary Change* 50 (2006): 63–82.

Appendix A. Impacts of Global Climate Change at Different Annual Mean Global Temperature Increases

ΔT (°C)[a]	Table A1. Projected global impacts resulting from changes in global average surface temperature relative to 1990
0.1	• Water: Glaciers retreating worldwide; changes in rate and seasonality of stream flows; increase in extreme rainfall patterns causing drought and flood. • Coasts: Sea level increasing at .18 m/year.[b] • Ecosystems: 80 percent of 143 studies of phenological, morphological, and distributional changes in species show changes in direction consistent with expected response to climate change, for example, advance of spring by five days, losses in alpine flora. • Abrupt events: Substantial and increasing damage due to extreme weather events partly caused by climatic factors, particularly affecting small islands.
0.5	• Water: 300 million to 1,600 million additional people suffer increased water stress depending on socioeconomic scenario and extent of global climate management. • Food: 18 million to 60 million additional people at risk of hunger; 20-million- to 35-million ton loss in cereal production, depending on socioeconomic scenario, global climate management (GCM—mitigation of and adaptation to climate change), and realization of CO_2 fertilization effect. • Health: Increase in heat waves and associated mortality, decrease in cold spells and associated mortality; increased risk of malaria and dengue. • Ecosystems: Oceans continue to acidify, with unknown consequences for entire marine ecosystem; 80 percent loss of coral reefs due to climate change—induced changes in water chemistry and bleaching; potential disruption of ecosystems as predators, prey, and pollinators respond at different rates to climatic changes; increase in damage from pests and fires; 10 percent of ecosystems transformed, variously losing between 2 to 47 percent of their extent; loss of cool conifer forest; further extinctions in cloud forests. • Abrupt events: Further increase in extreme precipitation causing drought, flood, landslides, likely to be exacerbated by more intense El Niño; rise in insurance prices and decreased availability of insurance.
1.0	• Coasts: Onset of Greenland Ice Sheet melt, causing eventual additional 23 feet (7meters) of sea level rise over several centuries.
1.5	• Water: 1.0 billion to 2.8 billion additional people experience increased water stress, depending on socioeconomic scenario and global climate management. • Food: Threshold above which agricultural yields fall in developed world; –12 million to +220 million additional people at risk of hunger and 30-million- to 180-million-ton loss in global cereal production, depending on socioeconomic scenario, global climate management, and realization of CO_2 fertilization effect. • Coasts: 12 million to 26 million people (less those protected by adaptation) displaced from coasts owing to sea level rise and cyclones. • Health: Additional millions at risk of malaria, particularly in Africa and Asia, depending on socioeconomic scenario. • Ecosystems: 97 percent loss of coral reefs;16 percent of global ecosystems transformed—ecosystems variously lose between 5 and 66 percent of their extent.
1.5–2.5	• Water: 0.9 billion to 3.5 billion additional people suffer increased water stress. • Ecosystems: Conversion of land vegetation from carbon sink to net carbon source; collapse of Amazon rainforest.
.5	• Water: 1.0 billion to 3.0 billion additional people suffer increased water stress, depending on socioeconomic scenario and global change management; irrigation requirements increase in 12 of 17 world regions.

Appendix A. Impacts of Global Climate Change at Different Annual Mean Global Temperature Increases (*continued*)

ΔT (°C)[a]	Table A1. Projected global impacts resulting from changes in global average surface temperature relative to 1990
	• Food: –20 million (minus figure indicates low probability of small improvement) to +400 million additional people at risk of hunger and 20-million- to 400-million-ton loss of global cereal production, depending on socioeconomic scenario and realization of CO_2 fertilization effect; 65 countries lose 16 percent of agricultural GDP even if CO_2 fertilization assumed to occur. • Coasts: 25 milion to 40 million additional people displaced from coasts owing to sea level rise (less those protected by adaptation). • Health: 17 to 18 percent increase in seasonal and perennial potential malarial transmission zones, exposing 200 million to 300 million additional people; overall increase for all zones 10 percent; 50 to 60 percent of world population exposed to dengue compared to 30 percent in 1990. • Ecosystems: Few ecosystems can adapt; 50 percent of nature reserves cannot fulfill conservation objectives, 22 percent of ecosystems are transformed, 22 percent loss of coastal wetlands; ecosystems variously lose between 7 and 74 percent of their extent.
3.5	• Food: Entire regions are out of agricultural production; –30 million to +600 additional million at risk of hunger. • Health: 25 percent increase in potential malarious zones; 40 percent increase in seasonal zones; 20 percent decrease in perennial zones. • Ecosystems: 44 percent loss of taiga, 60 percent loss of tundra; timber production increases by 17 percent. • Abrupt events: Probability of MOC shutdown = 50 percent according to many experts.[c]
Africa	
0.1	Abrupt change in regional rainfall causes drought and water stress, food insecurity and loss of grassland in the Sahel.
0.5	Decreases in crop yields such as barley and rice estimated at 10 percent; significant loss of Karoo, the richest floral area in the world; increased risk of death due to flooding; southern Kalahari dunefield begins to activate.
1.5	Large-scale displacement of people (climate refugees from low food security, poverty, and water stress) in Maghreb as rainfall declines by at least 40 percent; Kalahari dunefields begin to lose vegetation and destabilize structurally.
1.5–2.5	80 percent of Karoo lost, endangering 2,800 plant species with extinction; loss of Fynbos, causing extinction of endemics; 5 southern African parks lose more than 40 percent of animals; Great Lakes wetland ecosystems collapse; fisheries lost in Malawi; crop failures of 75 percent in southern Africa; all Kalahari dunefields may be mobile, threatening sub-Saharan ecosystems and agriculture with inundation by sand.
2.5	70 to 80 percent of additional millions of people at risk from hunger are in Africa.
3.5	70 to 80 percent of additional millions of people at risk from hunger are in Africa.
1.5–4.0	Crop failure rises by 50 to 75 percent in southern Africa.
Americas	
0.1	Extinction of Golden Toad in Central America.
0.5	Serious drinking water, energy, and agricultural problems in Peru following glacier melt; increased risk of death due to flooding; increased crop yields in North America in areas not affected by drought if CO_2 fertilization occurs.

(*continued*)

Appendix A. Impacts of Global Climate Change at Different Annual Mean Global Temperature Increases (*continued*)

ΔT(°C)	Table A2. Projected regional impacts resulting from changes in global average surface temperature relative to 1990
1.5	Vector-borne disease expands poleward; thus, 50 percent increase in malarial risk in North America; extinction of many Hawaiian endemic birds and impacts on salmonid fish.
1.5–2.5	Maple trees threatened in North American temperate forest.
2.5	50 percent loss of world's most productive duck habitat; large loss of migratory bird habitat.
Antarctic	
0.1	Collapse of ice shelves; changes in penguin populations.
1.5	Potential ecosystem disruption due to extinction of key molluscs.
1.5–4.0	Potential to trigger melting of the West Antarctic Ice Sheet, raising sea levels by a further 5 to 6 meters, or .60 to 1.20 m/century.
Arctic	
0.1	Local temperature rise of 1.8°C; damage to built infrastructures due to thawing permafrost; accelerating sea ice loss now at 360,000 square kilometers per decade.
0.5	Only 53 percent of wooded tundra remains stable.
1.5	Destruction of Inuit hunting culture; total loss of summer Arctic sea ice; likely extinction of polar bear, walrus; disruption of ecosystem due to 60 percent lemming decline; only 42 percent existing Arctic tundra remains stable, high arctic breeding shorebirds and geese in danger, common mid-Arctic species also impacted.
Asia 1.5	1.8 billion to 4.2 billion people experience decrease in water stress (depending on socioeconomic scenario and global change management model used) but largely in wet season and not in arid areas; vector-borne disease increases poleward; 50 percent loss of Chinese boreal forest; 50 percent loss of Sundarbans wetlands in Bangladesh.
1.5–2.5	Large impacts (desertification, permafrost shift) on Tibetan Plateau; complete loss of Chinese boreal forest, food production threatened in south.
2.5	Chinese rice yields fall by 10 to 20 percent or increase by 10 to 20 percent if CO_2 fertilization is realized.
Australia	
0.5	Extinctions in Dryandra forest; 50 percent loss of Queensland rainforest, endangering endemic frogs and reptiles.
1.5	Risk of extinctions accelerates in northern Australia, such as golden bowerbird; 50 percent loss of Kakadu wetland.
1.5–2.5	Total loss of Kakadu wetlands and Alpine zone.
2.5	50 percent loss of eucalypts; 24 percent loss of suitable (and 80 percent loss of original) range of endemic butterflies.
3.5	Out of agricultural production; total loss of alpine zone.
Europe and Russia	
0.1	Northward shifts in plankton distribution in North Sea, likely to have caused observed decline in sand eels and hence breeding failure of seabirds; changes in fish distributions; extreme heat and drought in 2003 that caused 25,000 human deaths has been attributed to anthropogenic climate change.

Appendix A. Impacts of Global Climate Change at Different Annual Mean Global Temperature Increases (*continued*)

ΔT(°C)	Table A2. Projected regional impacts resulting from changes in global average surface temperature relative to 1990
0.5	Increased crop yields if CO_2 fertilization occurs in areas not affected by drought; increased drought in steppes and around the Mediterranean, causing water stress and crop failure.
1.5	Tripling of bad harvests in spite of increased crop yields during good harvests.
2.5	Alpine species near extinction; 60 percent of species lost from Mediterranean region; high fire risk in Mediterranean region; large loss of migratory bird habitat.
3.5	38 percent of European alpine species lose 90 percent of range; in Russia, 5 to 12 percent drop in production, including 14 to 41 percent in agricultural regions.

Source: Rachel Warren, "Impacts of Global Climate Change at Different Annual Mean Global Temperature Increases," in *Avoiding Dangerous Climate Change,* edited by H. J. Schellnhuber and others (Cambridge University Press, 2006).

a. For Fahrenheit equivalent, multiply by 1.8.

b. For equivalent in feet, multiply number of meters by 3.28.

c. Expert judgments and models predict increasing probability of complete MOC collapse in range of 1 to 5°C temperature increase; predictions of 50 percent collapse probability range from 2 to 5°C.

Appendix B. Coastal Cities and Agricultural Deltas with at Least Part of Their Area Less Than 2 Meters (6.6 feet) Elevation above Present Sea Level

These areas are deemed to be potentially vulnerable to sea level rise as a result of low elevation combined with large human populations and importance to global economics or food security.

City or delta	Population	Source	Low point (meters)	High point (meters)	Source
United States and Canada					
New Orleans (before Hurricane Katrina)	454,863	a	–2.4	8	d
Galveston	57,466	a	0.0	2	c
Beaumont, Texas	111,799	a	0.0	8	c
Miami	386,417	a	0.0	9	d
Jacksonville, Florida	782,623	a	0.0	12	d
Sacramento, California	456,441	a	0.0	20	c
Corpus Christi	283,474	a	0.0	20	c
Houston	2,016,582	a	0.0	26	d
Mobile	191,544	a	0.0	63	c
Newark, New Jersey	280,666	a	0.0	86	d
Boston	559,034	a	0.0	102	d
Long Beach, California	474,014	a	0.0	111	d
New York	8,143,197	a	0.0	126	d
Washington, D.C.	550,521	a	0.3	126	d
Philadelphia	1,463,281	a	0.0	136	d
Vancouver	578,041	a	0.0	150	c
Baltimore	635,815	a	0.0	150	d
Seattle	573,911	a	0.0	160	d
San Diego	1,255,540	a	0.0	253	d

(continued)

Appendix B. Coastal Cities and Agricultural Deltas with at Least Part of Their Area Less Than 2 Meters (6.6 feet) Elevation above Present Sea Level (*continued*)

City or delta	Population	Source	Low point (meters)	High point (meters)	Source
San Francisco	739,426	a	0.0	287	d
Portland, Oregon	533,427	a	0.0	330	d
Oakland, California	395,274	a	0.0	540	d
San Jose, California	912,332	a	0.0	654	d
Honolulu	377,379	a	0.0	1,237	d
Los Angeles	3,844,829	a	0.0	1,561	d
Mexico					
Cancún	526,701	a	0.0	13	c
Cozumel	71,401	a	0.0	15	c
Vera Cruz	444,438	a	0.0	35	c
Ixtapa	290,076	a	0.0	170	c
South America					
Guayaquil, Ecuador	1,985,379	a	0.0	12	c
Santos, Brazil	416,100	a	0.0	13	c
Pelotas, Brazil	319,100	a	0.0	17	c
Bahia Blanca, Argentina	274,509	a	0.0	50	c
Porto Alegre, Brazil	1,386,900	a	0.0	100	c
Rio de Janeiro	6,094,200	a	0.0	320	c
Australia and New Zealand					
Darwin	69,354	a	0.0	30	c
Cairns	125,132	a	0.0	35	c
Melbourne	61,670	a	0.0	50	c
Auckland	416,900	a	0.0	65	c
Newcastle	145,633	a	0.0	100	c
Sydney	146,297	a	0.0	107	c
Adelaide	14,361	a	0.0	350	c
Asia					
Alor Star, Malaysia	186,524	a	0.0	8	c
Shanghai	14,230,992	a	0.0	8	c
Barisal, Bangladesh	8,514,000	a	0.0	10	c
Ho Chi Minh City	5,059,800	a	0.0	10	c
Sinuiju, North Korea	326,011	a	0.0	10	c
Kolkata	13,211,853	a	0.0	15	c
Dandong, China	684,190	a	0.0	20	c
Dhaka	10,403,597	a	2.0	22	c
Bangkok	6,355,144	a	2.0	25	c
Singapore	4,163,700	a	0.0	30	c
Pondicherry, Andhra Pradesh, India	505,959	a	0.0	50	c
Karachi	9,339,023	a	0.0	85	c
Kota Bharu, Malaysia	252,714	a	0.0	600	c
North Africa					
Port Said	469,533	a	0.0	15	c
Alexandria	3,328,196	a	0.0	17	c
Banghazi	446,250	a	0.0	25	c
Tripoli	1,150,000	a	0.0	40	c
Rabat	1,622,860	a	0.0	67	c

Appendix B. Coastal Cities and Agricultural Deltas with at Least Part of Their Area Less Than 2 Meters (6.6 feet) Elevation above Present Sea Level (*continued*)

City or delta	Population	Source	Low point (meters)	High point (meters)	Source
Tunis	728,453	a	0.0	100	c
Casablanca	2,933,684	a	0.0	140	c
Algiers	1,519,570	a	0.0	360	c
West Africa					
Banjul, The Gambia	351,018	a	0.0	10	c
Port-Gentil, Gabon	79,225	a	0.0	10	c
Lagos	5,195,247	a	0.0	18	c
Douala, Cameroon	1,494,700	a	0.0	40	c
Dakar	2,564,900	a	0.0	46	c
Lomé, Togo	1,194,000	a	0.0	50	c
Accra, Ghana	1,658,937	a	0.0	62	c
Conakry, Guinea	1,092,936	a	0.0	266	c
Abidjan, Ivory Coast	1,934,342	a	2.0	90	c
East Africa					
Quelimane, Mozambique	153,187	a	0.0	10	c
Mombasa, Kenya	828,500	a	0.0	25	c
Dar es Salaam, Tanzania	2,336,055	a	0.0	40	c
Tanga and Pemba, Tanzania	179,400	a	0.0	40	c
Maputo, Mozambique	989,386	a	0.0	55	c
Zanzibar, Tanzania	205,870	a	0.0	75	c
Europe					
Amsterdam	743,070	a	0.0	5	c
Venice	269,780	a	0.0	11	c
The Hague	475,627	a	0.0	15	c
Copenhagen	1,084,885	a	0.0	35	c
Dublin	506,211	a	0.0	40	c
London	7,172,091	a	0.0	50	c
Galway	72,414	a	0.0	50	c
Montpelier	244,300	a	0.0	55	c
Arhus, Denmark	228,674	a	0.0	75	c
Odens, Denmark	152,060	a	2.0	30	c
Mega-deltas					
Mekong Delta	20,200,000	b	0.0	5	c
Yellow River Delta	614,000	b	0.0	5	c
Mississippi Delta	1,790,000	b	0.0	5	c
Red River Delta	5,710,000	b	0.0	6	c
Ganges-Brahmaputra	111,000,000	b	0.0	10	c
Chao Phraya Delta	13,700,000	b	0.0	11	c
Yangtze Delta	42,100,000	b	0.0	14	c
Nile Delta	47,800,000	b	0.0	20	c
Niger Delta	3,730,000	b	0.0	20	c

Sources: a. T. Brinkhoff, "The Principal Agglomerations of the World," 2007 (www.citypopulation.de). b. J. P. Ericson and others, "Effective Sea-Level Rise and Deltas: Causes of Change and Human Dimension Implications," *Global and Planetary Change* 50 (2006): 63–82. c. Google, Google Earth, 4.1 ed. (Mountain View, Calif.: Google, 2007). d. United States Geological Survey, "Elevations and Distances in the United States," online edition (Reston, Va.: 2005) (http://erg.usgs.gov/isb/pubs/booklets/elvadist/elvadist.html).

Notes

1. T. R. Carter and others, "New Assessment Methods and the Characterisation of Future Conditions," in *Climate Change 2007: Impacts, Adaptation and Vulnerability; Contribution of Working Group II to the Fourth Assessment Report of the Intergovernmental Panel on Climate Change,* edited by M. L. Parry and others (Cambridge University Press, 2007), pp. 133–71.

2. IPCC, "Summary for Policymakers," in *IPCC Special Report: Emissions Scenarios* (Geneva: 2000).

3. Carter and others, "New Assessment Methods and the Characterisation of Future Conditions."

4. Michael MacCracken, "Prediction versus Projection—Forecast versus Possibility," *WeatherZine,* no. 26 (2001): 3–4 (http://sciencepolicy.colorado.edu/zine/archives/1-29/26/guest.html).

5. IPCC, "Summary for Policymakers," in *IPCC Special Report: Emissions Scenarios.*

6. IPCC, "Summary for Policymakers," in *Climate Change 2007: The Physical Science Basis; Contribution of Working Group I to the Fourth Assessment Report of the Intergovernmental Panel on Climate Change,* edited by Susan Solomon and others (Cambridge University Press, 2007).

7. IPCC, "Summary for Policymakers," in *Climate Change 2007: Mitigation of Climate Change; Contribution of Working Group III to the Fourth Assessment Report of the Intergovernmental Panel on Climate Change,* edited by B. Metz and others (Cambridge University Press, 2007).

8. J. E. Hansen, "Scientific Reticence and Sea Level Rise," *Environmental Research Letters* 2 (2007): 024002, doi (digital object identifier): 10.1088/1748-9326/2/2/024002; G. J. Holland and P. J. Webster, "Heightened Tropical Cyclone Activity in the North Atlantic: Natural Variability or Climate Trend?" *Proceedings of the Royal Society,* series A 365 (2007): 2695–2716, doi: 10.1098/rsta.2007.2083; G. A. Meehl and others, "Global Climate Projections," in *Climate Change 2007: The Physical Science Basis,* pp. 747–846; J. T. Overpeck and others, "Paleoclimatic Evidence for Future Ice-Sheet Instability and Rapid Sea-Level Rise," *Science* 311 (2006): 1747–50; B. A. Pittock, "Are Scientists Underestimating Climate Change?" *Eos: Transactions of the American Geophysical Union* 34 (2006): 340–41; Stefan Rahmstorf and others, "Recent Climate Observations Compared to Projections," *Science* 316 (2007): 709; M. E. Schlesinger and others, "Assessing the Risk of a Collapse of the Atlantic Thermohaline Circulation," in *Avoiding Dangerous Climate Change,* edited by H. J. Schellnhuber and others (Cambridge University Press, 2006); J. Stroeve and others, "Arctic Sea Ice Decline: Faster Than Forecast," *Geophysical Research Letters* (2007): L09501, doi: 10.1029/2007GL029703; K. Trenberth, K. E. Shea, and D. J. Shea, "Atlantic Hurricanes and Natural Variability in 2005," *Geophysical Research Letters* 33 (2006): L12704, doi: 10.1029/2006GL026894; F. J. Wentz and others, "How Much More Rain Will Global Warming Bring?" *Science* 317 (2007): 233–235, doi: 10.1126/science.1140746.

9. Hansen, "Scientific Reticence and Sea Level Rise"; Pittock, "Are Scientists Underestimating Climate Change?"; Rahmstorf and others, "Recent Climate Observations Compared to Projections," p. 709; Stroeve and others, "Arctic Sea Ice Decline"; Wentz and others, "How Much More Rain Will Global Warming Bring?"

10. Pittock, "Are Scientists Underestimating Climate Change?"; J. G. Canadell and others, "Contributions to Accelerating Atmospheric CO_2 Growth from Economic Activity, Carbon Intensity, and Efficiency of Natural Sinks," *Proceedings of the National Academy of Sciences USA* 104 (2007): 18866–70, doi: 10.1073/pnas.0702737104.

11. Stefan Rahmstorf, "A Semi-Empirical Approach to Projecting Future Sea-Level Rise," *Science* 315 (2007): 368–70, doi: 10.1126/science.1135456; Stroeve and others, "Arctic Sea Ice Decline"; Wentz and others, "How Much More Rain Will Global Warming Bring?"

12. J. A. Kettleborough, B. B. B. Booth, P. A. Stott, and M. R. Allen, "Estimates of Uncertainty in Predictions of Global Mean Surface Temperature," *Journal of Climate* 20 (2007): 843–55.

13. Meehl and others, "Global Climate Projections."

14. IPCC, *Climate Change 2007: Impacts, Adaptation and Vulnerability;* R. Warren, "Impacts of Global Climate Change at Different Annual Mean Global Temperature Increases," in *Avoiding Dangerous Climate Change,* edited by Schellnhuber and others.

15. G. McGranahan, D. Balk, and B. Anderson, "The Rising Tide: Assessing the Risks of Climate Change and Human Settlements in Low Elevation Coastal Zones," *Environment & Urbanization* 19 (2007): 17–37, doi: 10.1177/0956247807076960.

16. IPCC, "Summary for Policymakers," in *Climate Change 2007: The Physical Science Basis.*

17. P. Lemke and others, "Observations: Changes in Snow, Ice and Frozen Ground," in *Climate Change 2007: The Physical Science Basis.*

18. C. J. Van der Veen, "Fracture Propagation as a Means of Rapidly Transferring Surface Meltwater to the Base of Glaciers," *Geophysical Research Letters* 34 (2007): L01501, doi: 10.1029/2006GL028385; H. J. Zwally and others, "Surface Melt–Induced Acceleration of Greenland Ice-Sheet Flow," *Science* 297 (2002): 218–22.

19. IPCC, "Summary for Policymakers," in *Climate Change 2007: The Physical Science Basis.*

20. Rahmstorf and others, "Recent Climate Observations Compared to Projections."

21. IPCC, "Summary for Policymakers," in *Climate Change 2007: The Physical Science Basis.*

22. The experts interviewed were Richard Alley, John Church, Jonathan Gregory, John Hunter, Michael MacCracken, R. Steven Nerem, Michael Oppenheimer, Jonathan Overpeck, and Konrad Steffen.

23. J. G. Titus and V. Narayanan, "The Risk of Sea Level Rise," *Climatic Change* 33 (1996): 151–212; D. G. Vaughan and J. R. Spouge, "Risk Estimation of Collapse of the

West Antarctic Ice Sheet," *Climatic Change* 52 (2002): 65–91; K. Zickfeld and others, "Expert Judgements on the Response of the Atlantic Meridional Overturning Circulation to Climate Change," *Climatic Change* 82 (2007): 235–65.

24. Rahmstorf, "A Semi-Empirical Approach to Projecting Future Sea-Level Rise."

25. Ibid.; J. E. Hansen, "Scientific Reticence and Sea Level Rise."

26. J. Hansen and others, "Climate Change and Trace Gases," *Philosophical Transactions of the Royal Society* 365 (2007): 1925–54, doi: 10.1098/rsta.2007.2052.

27. Ibid.

28. Overpeck and others, "Paleoclimatic Evidence for Future Ice-Sheet Instability and Rapid Sea-Level Rise."

29. E. J. Rohling, K. Grant, C. Hemleben, M. Siddall, B. A. A. Hoogakker, M. Bolshaw, and M. Kucera, "High Rates of Sea-Level Rise During the Last Interglacial Period," *Nature Geoscience* 1 (2008): 38–42.

30. IPCC, "Summary for Policymakers," in *Climate Change 2001: The Scientific Basis, Contribution of Working Group I to the Third Assessment Report of the Intergovernmental Panel on Climate Change,* edited by J. T. Houghton and others (Cambridge University Press, 2001).

31. Ibid.

32. IPCC, "Summary for Policymakers," in *Climate Change 2007: The Physical Science Basis;* Overpeck and others, "Paleoclimatic Evidence for Future Ice-Sheet Instability and Rapid Sea-Level Rise."

33. IPCC, "Summary for Policymakers," in *Climate Change 2007: Impacts, Adaptation and Vulnerability.*

34. Meehl and others, "Global Climate Projections." J. H. Christensen and others, "Regional Climate Projections," in *Climate Change 2007: The Physical Science Basis,* pp. 847–940.

35. O. A. Anisimov and others, "Polar Regions (Arctic and Antarctic)," in *Climate Change 2007: Impacts, Adaptation and Vulnerability,* pp. 653–86.

36. Meehl and others, "Global Climate Projections"; R. Seager and others, "Model Projections of an Imminent Transition to a More Arid Climate in Southwestern North America," *Science* 316 (2007): 1181–84, doi: 10.1126/science.1139601.

37. A. Fischlin and others, "Ecosystems: Their Properties, Goods, and Services," in *Climate Change 2007: Impacts, Adaptation and Vulnerability,* pp. 211–72.

38. Compare to figure 5.1a in W. E. Easterling and others. 2007. "Food, Fibre and Forest Products." In *Climate Change 2007: Impacts, Adaptation and Vulnerability,* pp. 273–313.

39. Fischlin and others, "Ecosystems: Their Properties, Goods, and Services," pp. 211–72.

40. M. B. Baettig, M. Wild, and D. M. Imboden, "Climate Change Index: Where Climate Change May Be Most Prominent in the 21st Century," *Geophysical Research Letters* 34 (2007): L01705, doi: 10.1029/2006GL028159.

41. Fischlin and others, "Ecosystems: Their Properties, Goods, and Services."

42. IPCC, "Summary for Policymakers," in *Climate Change 2007: The Physical Science Basis,* pp. 1–18.

43. IPCC, "Summary for Policymakers," in *Climate Change 2007: Impacts, Adaptation and Vulnerability,* pp. 7–22.

44. R. McTaggart-Cowan and others, "Analysis of Hurricane Catarina (2004)," *Monthly Weather Review* 134 (2006): 3029–53, doi: 10.1175/MWR3330.1.

45. J. L. Franklin, "Tropical Cyclone Report: Hurricane Vince, 8–11 October 2005," National Hurricane Center, 2006 (www.nhc.noaa.gov/pdf/TCR-AL242005_Vince.pdf).

46. S. Al-Nahdy and J. Krane, "Thousands Evacuated Ahead of Cyclone," *Miami Herald,* June 6, 2007.

47. IPCC, "Summary for Policymakers," in *Climate Change 2007: The Physical Science Basis.*

48. Ibid.

49. M. E. Schlesinger and others, "Assessing the Risk of a Collapse of the Atlantic Thermohaline Circulation," in *Avoiding Dangerous Climate Change,* edited by Schellnhuber and others; Zickfeld and others, "Expert Judgements on the Response of the Atlantic Meridional Overturning Circulation to Climate Change."

50. R. Seager and others, "Is the Gulf Stream Responsible for Europe's Mild Winters?" *Quarterly Journal of the Royal Meteorological Society* 128 (2002): 2563–86.

51. M. Hulme, "Abrupt Climate Change: Can Society Cope?" *Philosophical Transactions of the Royal Society, London* (series A) 361 (2003): 2001–21.

52. Nigel W. Arnell, "Global Impacts of Abrupt Climate Change: An Initial Assessment," working paper 99 (Norwich, England: Tyndall Centre for Climate Change Research, 2006).

53. Meehl and others, "Global Climate Projections."

54. Arnell, "Global Impacts of Abrupt Climate Change."

55. Ibid.

56. Ibid.

57. Ibid.

58. A. Levermann and others, "Dynamic Sea Level Changes Following Changes in the Thermohaline Circulation," *Climate Dynamics* 24 (2005): 347–54.

59. Zickfeld and others, "Expert Judgements on the Response of the Atlantic Meridional Overturning Circulation to Climate Change."

60. R. A. Wood, M. Vellinga, and R. Thorpe, "Global Warming and Thermohaline Circulation Stability," *Philosophical Transactions of the Royal Society, London* (series A) 361 (2003): 1961–75.

61. IPCC, "Summary for Policymakers," in *Climate Change 2007: The Physical Science Basis.*

62. Levermann and others, "Dynamic Sea Level Changes Following Changes in the Thermohaline Circulation"; Zickfeld and others, "Expert Judgements on the Response of the Atlantic Meridional Overturning Circulation to Climate Change."

63. C. Rosenzweig and W. D. Solecki, eds., *U.S National Assessment of the Potential Consequences of Climate Variability and Change* (New York: U.S. Global Change Research Program, 2001).

64. Meehl and others, "Global Climate Projections."

65. Nicholls and others, "Coastal Systems and Low-Lying Areas"; C. Small, V. Gornitz, and J. E. Cohen, "Coastal Hazards and the Global Distribution of Human Population," *Environmental Geosciences* 7 (2000): 3–12.

66. G. McGranahan, D. Balk, and B. Anderson, "The Rising Tide: Assessing the Risks of Climate Change and Human Settlements in Low Elevation Coastal Zones," *Environment & Urbanization* 19 (2007): 17–37, doi: 10.1177/0956247807076960.

67. Nicholls and others, "Coastal Systems and Low-Lying Areas."

68. Ibid.

69. Ibid.

70. Nicholls, "Coastal Flooding and Wetland Loss in the 21st Century."

71. T. E. Dahl, "Status and Trends of Wetlands in the Conterminous United States 1998 to 2004" (U.S. Department of the Interior, Fish and Wildlife Service, 2006), p. 112.

72. Nicholls and others, "Coastal Systems and Low-Lying Areas"; Nicholls, "Coastal Flooding and Wetland Loss in the 21st Century."

73. IPCC, "Summary for Policymakers," in *Climate Change 2007: The Physical Science Basis.*

74. Nicholls and others, "Coastal Systems and Low-Lying Areas."

75. Ibid.; Rosenzweig and Solecki, *U.S National Assessment of the Potential Consequences of Climate Variability and Change.*

76. Anu Mittal (United States Government Accountability Office), "Army Corps of Engineers: History of the Lake Pontchartrain and Vicinity Hurricane Protection Project," testimony before the Committee on Environment and Public Works, U.S. Senate, November 9, 2005, p. 1. Available at http://www.gao.gov/new.items/d06244t.pdf.

77. V. Gornitz, R. Horton, A. Siebert, and C. Rosenzweig, "Vulnerability of New York City to Storms and Sea Level Rise," *Geological Society of America Abstracts with Programs* 38 (2006): 335 (serially published abstracts: http://gsa.confex.com/gsa/2006AM/finalprogram/abstract_112613.htm); Rosenzweig and Solecki, *U.S National Assessment of the Potential Consequences of Climate Variability and Change.*

78. Ibid.

79. Meehl and others, "Global Climate Projections"; Nicholls and others, "Coastal Systems and Low-Lying Areas."

four

Security Implications of Climate Scenario 1

Expected Climate Change over the Next Thirty Years

John Podesta and Peter Ogden

The effects of climate change projected in this chapter are based on the A1B greenhouse gas emission scenario of the *Fourth Assessment Report of the Intergovernmental Panel on Climate Change.*[1] It is a scenario in which people and nations are threatened by massive food and water shortages, devastating natural disasters, and deadly disease outbreaks. It is also to a large extent inevitable.

Scenario Overview: Expected Climate Change

There is no foreseeable political or technological solution that will enable us to avert many of the climatic impacts projected in this chapter. The world will confront elements of this climate change scenario even if, for instance, the United States were to enter into an international carbon cap and trade system in the near future. The scientific community, meanwhile, remains far from a technological breakthrough that would lead to a decisive, near-term reduction in the concentration of carbon dioxide in the atmosphere.

Furthermore, the Intergovernmental Panel on Climate Change's (IPCC) A1B greenhouse gas emission scenario assumes that climate change will not trigger any significant positive feedback loops (for example, the release of carbon dioxide and methane from thawing permafrost). Such feedback loops would multiply and magnify the impacts of climate change, creating an even more hostile environment than the one projected here.

Thus, it is not alarmist to say that this scenario may be the best we can hope for. It is certainly the least we ought to prepare for. To do so, we must recognize that the foreign policy and national security implications of climate change are as much determined by local political, social, and economic factors as by the magnitude of the climatic shift itself. As a rule, wealthier countries, and wealthier individuals, will be better able to adapt to the impacts of climate change, while the disadvantaged will suffer the most. An increase in rainfall, for instance, can be a blessing for a country that has the ability to capture, store, and distribute the additional water; however, it is a deadly source of soil erosion for a country that does not have adequate land management practices or infrastructure.[2]

Consequently, even though the IPCC projects that the temperature increases at higher latitudes will be approximately twice the global average, it will be the developing nations in the Earth's low latitudinal bands that will be most adversely affected by climate change. In the developing world, even a relatively small climatic shift can trigger or exacerbate food shortages, water scarcity, destructive weather events, the spread of disease, human migration, and natural resource competition. These crises are all the more dangerous because they are interwoven and self-perpetuating: water shortages can lead to food shortages, which can lead to conflict over remaining resources, which can drive human migration, which, in turn, can create new food shortages in new regions.

Once under way, this chain reaction becomes increasingly difficult to stop, and therefore it is critical that policymakers do all they can to prevent that first climate change domino—whether it be food scarcity or the outbreak of disease—from toppling. In this scenario, we identify each of the most threatening first dominos, where they are situated, and their cascading geopolitical implications.

Regional Sensitivity to Climate Change

The United States, like most wealthy and technologically advanced countries, will not experience destabilizing levels of internal migration as a result of climate change, but it will be affected. According to the IPCC, tropical cyclones will become increasingly intense in the coming decades, which will force the resettlement of people from coastal areas in the United States. This can have significant economic and political consequences, as was the case with the evacuation and permanent relocation of many Gulf Coast residents in the aftermath of Hurricane Katrina.[3]

In addition, the United States will need to contend with increased border stress as a result of the severe effects of climate change in parts of Mexico and the Caribbean. Northern Mexico will experience severe water shortages, which will drive immigration into the United States in spite of the increasingly treacherous border terrain. Likewise, the damage caused by storms and rising sea levels in the coastal areas of the Caribbean islands—where 60 percent of the Caribbean population lives—will increase the flow of immigrants from the region and generate political tension.[4]

It is in the developing world, however, where the impact of climate-induced migration will be most pronounced. Migration will widen the wealth gap between and within many of these countries. It will deprive developing countries of sorely needed economic and intellectual capital as the business and educated elite who have the means to emigrate abroad do so in greater numbers than ever before.[5] In some cases, it will even spark war by heightening competition over scarce resources and upsetting the cultural or ethnic order within a country or region.[6]

The three regions in which climate-induced migration will present the greatest geopolitical challenges are South Asia, Africa, and Europe.

South Asia

No region is more directly threatened by climate change–induced human migration than South Asia. Among the numerous threats that climate change will present to this region, the IPCC warns that "coastal areas, especially heavily populated mega-delta regions in South, East, and Southeast Asia, will be at greatest risk due to increased flooding from the sea and, in some mega-deltas, flooding from the rivers."[7]

Bangladesh, in particular, will be severely impacted by devastating floods, monsoons, melting glaciers, and tropical cyclones, as well as by the water contamination and ecosystem destruction caused by rising sea levels. Yet, even as the impact of climate change and other environmental factors will steadily render the low-lying regions of the country uninhabitable, the population of Bangladesh, which stands at 142 million as of 2008, is anticipated to increase by approximately 100 million people during the next few decades.[8] Many of the displaced will move inland, which will foment instability as the resettled population competes for already scarce resources with the established residents. Some will seek to migrate abroad, creating heightened political tension as they travel not only to neighboring states in the region, but to Europe and other regions as well.

Already, the India-Bangladesh border is a site of significant political friction, as evidenced by the 3,381-kilometer (2,100-mile), 2.5-meter-high (8-foot-high) iron border fence that India is in the process of building.[9] This fence is being constructed at a time when there are numerous signs of rising Islamic extremism in Bangladesh. In the wake of the United States' invasion of Afghanistan, hundreds of Taliban and jihadists are said to have found safe haven in Bangladesh.[10] The combination of deteriorating socioeconomic conditions, radical Islamic political groups, and dire environmental insecurity brought on by climate change could prove a volatile mix, one with severe regional and potentially global consequences.[11]

India will struggle to cope with a surge of displaced people from Bangladesh, in addition to those who will arrive from the small islands in the Bay of Bengal that are being slowly swallowed by the rising sea. Approximately 4 million people inhabit these islands, and many of them will have to be accommodated on the mainland eventually.[12]

Notably, the issue of climate change has not generated the same degree of public concern or political activity in India as it has in many other countries. A recent global public opinion poll found that only 19 percent of Indians believe that global warming is a sufficiently "serious and pressing problem" to merit taking immediate action if it involves significant cost.[13] In contrast, 42 percent of Chinese and 43 percent of Americans support taking action under such circumstances.[14]

In the coming decades, however, there will likely be some shift in domestic attitudes—and in international expectations for greater engagement—as India's booming population begins to confront acute environmental stress stemming from climate change and environmental degradation.[15] According to the IPCC, agricultural productivity throughout South Asia will decrease significantly owing to high temperature, severe drought, flooding, and soil degradation. India will reach a state of water stress before 2025, and cases of waterborne infectious disease and cholera are projected to increase.[16]

To date India has steadfastly refused to commit itself to any binding limits on its carbon emissions, but there are indications that domestic policymakers are beginning to recognize the need to start facing the challenges posed by climate change. On July 13, 2007, Prime Minister Manmohan Singh chaired the inaugural meeting of the National Council on Climate Change to

Table 4-1. Annual Energy Consumption and Carbon Emissions in China, India, United States, and World, 2003 and 2030

Total energy use, 2003 actual/2030 projected, and percent annual growth rate

Consumption and emissions	China	India	United States	World
Total energy use (quadrillion BTUs)	45.5/139.1 4.2 %	14.0/32.5 3.2 %	98.1/133.9 1.2 %	420.7/721.6 2.0 %
CO_2 emissions (million metric tons)	3,541/10,716 4.2 %	1,023/2,205 2.9 %	5,796/8,115 1.3 %	11,878/26,180 2.1 %

Source: U.S. Energy Information Administration, *International Energy Outlook,* 2006.

oversee the development of India's first national climate change policy. What role India will play in international negotiations over carbon reduction policies in the years to come remains unclear, however, for its share of global energy demand and carbon emissions is projected to continue to lag far behind major consumers and emitters such as the United States and China for decades (see table 4-1).

Unfortunately, climate change is reducing the effectiveness of many of the development projects being financed by the international community in South Asia even as it is making them more necessary. The World Bank estimates that up to 40 percent of all overseas development assistance and concessional loans (loans that are provided to developing countries at low interest rates and with extended repayment periods) support projects that will be affected by climate change, yet few of these projects adequately account for the impacts of a shifting climate. As a result, dams are built on rivers that will dry up, and crops are planted in coastal areas that will be frequently flooded.[17] Furthermore, the water shortages brought about by climate change are coinciding with an increased tendency among donors and international financial organizations to promote the privatization of water, which frequently raises the cost for rural subsistence farmers to a level they cannot afford. This, in turn, foments tension between the poorer rural segments of society and the urban middle and upper classes by exacerbating existing economic and social inequities.

In Nepal, climate change is contributing to a phenomenon known as "glacial lake outburst," in which violent flood waves—reaching as high as 15 meters (50 feet)—destroy downstream settlements, dams, bridges, and other infrastructure. Millions of dollars in recent investment have been lost because hydropower and infrastructure design in Nepal largely fails to take these lethal floods into account. Ultimately, this puts further stress on the already beleaguered country as it struggles to preserve a fragile peace and reintegrate tens of thousands of Maoist insurgents. The stability of Nepal, which neighbors the entrenched conflict zone of Kashmir and the contested borders of China and India, has regional ramifications. An eruption of severe social or political turmoil could ripple across all of South Asia.

Nigeria and East Africa

The impact of climate change-induced migration will be felt throughout Africa, but its effects on Nigeria and East Africa pose particularly acute geopolitical challenges. This migration will be both internal and inter-national. The first domestic wave will likely be from agricultural regions to urban centers where more social services are available, which will impose a heavy burden on central governments. Simultaneously, the risk of state failure will increase as these migration patterns challenge the capacity of central governments to control stretches of their territory and their borders.

Nigeria, the most populous nation in Africa, will suffer from climate-induced drought, desertification, and sea level rise. Already, approximately 1,350 square miles of Nigerian land turns to desert each year, forcing both farmers and herdsmen to abandon their homes.[18] Lagos, the capital, is one of the West African coastal megacities that the IPCC identifies as at risk from sea level rise by 2015.[19] This, coupled with high projected population growth, will force significant migration and contribute to political and economic turmoil. It will, for instance, exacerbate the existing internal conflict over oil production in the Niger Delta.[20] To date, the Movement for the Emancipation of the Niger Delta (MEND) has carried out a successful campaign of armed attacks, sabotage, and kidnappings that has forced a shutdown of 25 percent of the country's oil output.[21] Given that Nigeria is the world's eighth largest oil exporter, this instability is having an impact on the price of oil, and it will have global strategic implications in the coming decades.[22] In addition to the Niger Delta issue, Nigeria must also contend with a Biafran separatist movement in its southeast.

The threat of regional conflagration, however, is highest in East Africa because of the concentration of weak or failing states, the numerous unresolved political disputes, and the severe impacts of climate change. Climate change will likely create large fluctuations in the amount of rainfall in East Africa during the next thirty years: a 5 to 20 percent increase in rainfall during the winter months will cause flooding and soil erosion, and a 5 to 10 percent decrease in the summer months will cause severe droughts.[23] This will jeopardize the livelihoods of millions of people and the economic capacity of the region. The agricultural sector constitutes some 40 percent of East Africa's GDP and provides a living for 80 percent of the population.[24]

In Darfur, water shortages have already led to the desertification of large tracts of farmland and grassland. The fierce competition that emerged between farmers and herdsmen over the remaining arable land combined with simmering ethnic and religious tensions to help ignite the first genocide of the twenty-first century.[25] This conflict has now spilled over into Chad and the Central African Republic. Meanwhile, the entire Horn of Africa continues to be threatened by a failed Somalia and other weak states. Al Qaeda cells are active in the region, and there is a danger that this area could become a central breeding ground and safe haven for jihadists as climate change pushes more states toward the brink of collapse.

Europe

Whereas most African and South Asian migration will be internal or regional, the expected decline in food production and fresh drinking water, combined with the increased conflict sparked by resource scarcity, will force more Africans and South Asians to migrate farther abroad.[26] This will likely result in a surge in the number of Muslim immigrants to the European Union (EU), which could exacerbate existing tensions and increase the likelihood of radicalization among members of Europe's growing (and often poorly assimilated) Islamic communities.

Already, the majority of immigrants to most Western European countries are Muslim. Muslims constitute approximately 5 percent of the European population, with the largest communities located in France, the Netherlands, Germany, and Denmark.[27] Europe's Muslim population is expected to double by 2025, and it will be much larger if the effects of climate change spur additional migration from Africa and South Asia.[28]

The degree of instability this generates will depend on how successfully these immigrant populations are integrated into European society. This process has not always gone well, as exemplified in 2005 by the riots in the poor and predominantly immigrant suburbs of Paris. The suspicion with which many view Europe's Muslim and immigrant communities has been intensified by "homegrown terrorist" attacks and plots, and the risk of a serious nationalist, anti-immigrant backlash is steadily increasing.

If the backlash is sufficiently severe, the European Union's (EU) cohesion will itself be tested. At present, the ease with which people can move between EU countries makes it extremely difficult to track or regulate both legal and illegal immigrants. In 2005, for instance, Spain granted amnesty to some 600,000 undocumented immigrants, and yet could provide few assurances that they would remain within Spain's borders.[29] The number of Africans

Figure 4-1. Key Migrant Routes from Africa to Europe

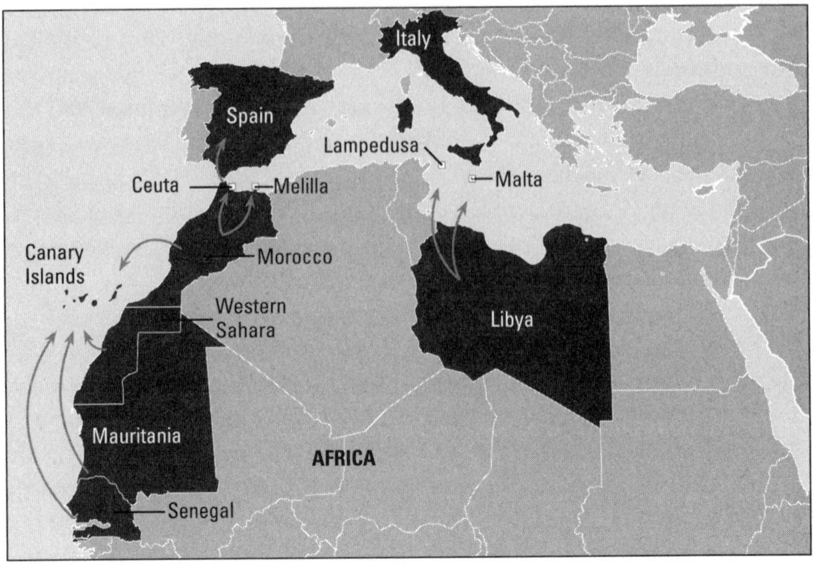

Source: BBC News (http://news.bbc.co.uk/2/hi/europe/5313560.stm).

who attempt to reach the Spanish Canary Islands—the southernmost European Union territory—has more than doubled since then. In 2006 at least 20,000 Africans attempted the perilous, often fatal, journey (see figure 4-1).[30]

To date, the EU has responded to this challenge with ad hoc measures, such as creating rapid-reaction border guard teams.[31] Although the influx of Muslim and other immigrants from Africa will continue to be viewed by some Europeans as a potential catalyst for economic growth at a time when the EU has a very low fertility rate, the viability of the EU's loose border controls will increasingly be called into question, and the absence of a common immigration policy will invariably lead to internal political tension. If a common immigration policy is not implemented, there is the possibility that significant border restrictions will reemerge, and this would slow the European Union's drive toward increased social, political, and economic integration.

Middle East and North Africa

Increasing water scarcity due to climate change will contribute to instability throughout the world. As we have discussed, in many parts of Africa populations will migrate in search of new water supplies, moving within and across borders and creating the conditions for social or political upheaval along the way. This was the case in Darfur, and its effects were felt throughout the entire region.

But water scarcity also shapes the geopolitical order when states engage in direct competition with neighbors over shrinking water supplies. This threat may evoke apocalyptic images of armies amassing in deserts to go to war over water, but the likelihood of such open conflict in this thirty-year scenario is low. There are a very limited number of situations in which it would make strategic sense for a country today to wage war in order to increase its water supply. Water does not have the economic value of a globally traded strategic commodity such as oil, and to reap significant benefit from a military operation would require capturing an entire watershed, cutting supply to the population currently dependent upon it, and then protecting the watershed and infrastructure from sabotage.[32]

Thus, we are not likely to see "water wars" per se. Rather, countries will more aggressively pursue the kinds of technological solutions and political arrangement that currently enable them to exist in regions that are stretched past their water limits.

This is likely to be the case in the Middle East, where water shortages will coincide with a population boom. The enormously intricate water politics of the region have been aptly described as a "hydropolitical security complex."[33] The Jordan River physically links the water interests of Syria, Lebanon, Jordan, Israel, and the Palestinian Authority; the Tigris and Euphrates rivers physically link the interests of Syria, Turkey, Iran, and Iraq. This hydrological environment is further complicated by the fact that 75 percent of all the water in the Middle East is located in Iran, Iraq, Syria, and Turkey.[34] Such conditions would be cause for political tension even in a region without a troubled history.

Turkey's regional position will likely be strengthened as a result of a deepening water crisis. As the point of origin of the Tigris and Euphrates rivers, Turkey is the only country in the Middle East that does not depend on water supplies whose headwaters are outside of its borders. Though Turkey is by no means a water-rich country, climate change per se will not significantly threaten its water supply within the next three decades.

But climate change will leave all of the other countries that are dependent on water from the Tigris and Euphrates rivers more vulnerable to deliberate supply disruption. Turkey is seeking to maximize this leverage with its massive Southeastern Anatolian Project, due for completion in 2010, which will give Turkey twenty-two dams and nineteen power plants along the Euphrates River, thereby reducing water supply farther downstream. The dams will also give Turkey the capacity to cut Syria's water supply by up to 40 percent and Iraq's water supply by up to 80 percent.[35]

Turkey's ability to use water as a political tool will become increasingly important in its relations with Syria. Turkey has previously threatened to cut off water in retaliation for Syria's support of the Kurdistan Workers Party, or PKK, and it has the capacity to reduce water supplies to Kurd-controlled northern Iraq.[36] Though Syria's support for the PKK ended in 1998, the chaos in Iraq could prompt an emboldened PKK to seek renewed support from potential regional allies.

Israel, already extremely water-poor, will only become more so. By 2025 Israel may have fewer than 500 cubic meters of water per capita per year, and by overpumping its existing water sources it is contributing to the gradual depletion and salinization of vital aquifers and rivers. Much of Israel's water, moreover, is located in politically fraught territory: one third of it flows from the Golan Heights and another third is in the mountain aquifer that straddles the West Bank and Israel.[37]

Israel will need to place additional importance on its relationship with Turkey, and a deeper alliance could be forged if a proposed water trading agreement—whereby Turkey would ship water directly to Israel in tankers—is eventually completed.[38] This new source of supply would not offset the added pressures of climate change and population growth, but it would deepen the countries' strategic ties and cushion any sudden short-term supply disruptions or embargoes.[39]

Israel's relations with Syria will also be strained by its need for the water resources of the Golan Heights. Although there is a mutual recognition that any peaceful and sustainable resolution over the Golan Heights will need to include a water-sharing agreement, Syria's interest in establishing permanent, direct access to the water resources of the Sea of Galilee through the Golan Heights will continue to complicate negotiations over the final demarcation of the border, as it did in 2000.

The region's water problems will be compounded by its population growth (see figure 4-2). According to current projections, the Middle Eastern and North African population could double in the next fifty years.[40] In the Middle East, the fastest-growing populations are in water-poor regions such as the Palestinian territories. In the West Bank, a lack of available freshwater has already contributed to food shortages and unemployment, and there have

Figure 4-2. Population of Middle East and North Africa by Age Group, 1950–2050

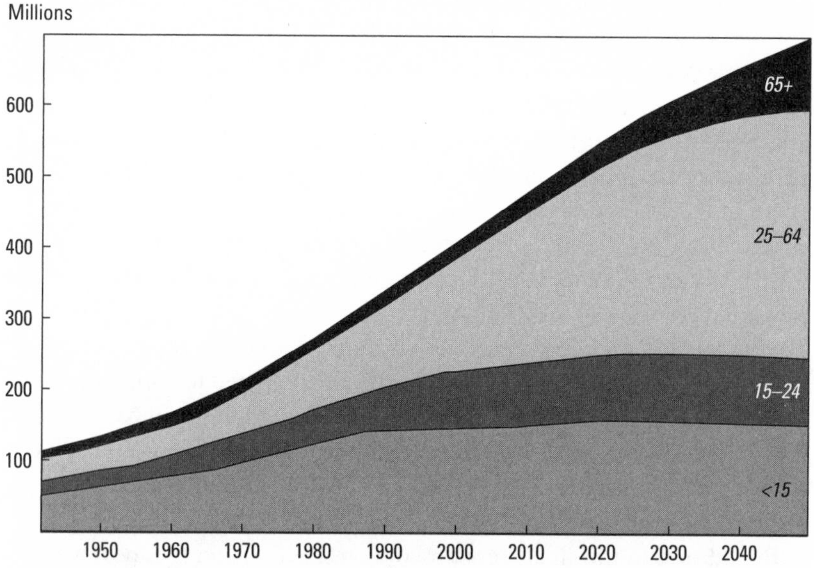

Source: United Nations Secretariat, Department of Economic and Social Affairs, Population Division, *World Population Prospects: The 2006 Revision* and *World Urbanization Prospects: The 2005 Revision* (http://esa.un.org/unpp).

been incidents of small, violent conflicts over water supplies.[41] These clashes will only become more prevalent as the population increases and available water resources diminish.

Disease

Climate change will have a range of decisively negative effects on global health during the next three decades, particularly in the developing world. The World Health Organization projects that the number of deaths linked to climate change could exceed 300,000 a year by 2030, and the total number of lost disease-adjusted life years (DALYS), a measurement that accounts also for injury and premature death, may surge to more than 11 million annually.[42]

The ways countries respond—or fail to respond—to these health challenges will have a significant impact on the geopolitical landscape. Waterborne and vector-borne diseases such as malaria and dengue fever will be particularly prevalent in countries that experience significant additional rainfall as a result of climate change.[43] Shortages of food or fresh drinking water will also render human populations both more susceptible to illness and less capable of rapidly recovering. Moreover, the risk of a pandemic is heightened when deteriorating conditions prompt human migration.[44]

This increase in the incidence of disease will inevitably generate disputes between nations over the movement of people. Immigrants, or even simply visitors, from a country in which there has been a significant disease outbreak may not be welcomed and could be subject to quarantine. If the policies that underlie such practices are perceived as discriminatory or motivated by factors other than legitimate health concerns, it will severely damage political relations.

This outcome might be averted if countries establish in advance common immigration policies that are specifically designed to cope with international health crises. However, it is most likely that this kind of coordination will occur only after the fact when the political damage has already been done, as was the case in Europe following several cholera epidemics in the mid-nineteenth century.[45]

In addition to the challenges posed by restrictions on the movement of people, restrictions on the movement of goods will be a source of economic and political turmoil. Countries affected by disease outbreaks could lose significant revenue from a decline in exports owing to limits or bans placed on products that originate there or are transported through them. The restrictions placed on India during a plague outbreak that lasted for seven weeks in 1994 cost it approximately $2 billion in trade revenue.[46] Countries that depend on tourism could be economically devastated by even relatively small outbreaks: the fear of severe acute respiratory syndrome (SARS) sharply curtailed international travel to Thailand in 2003, whereas the 2006 military coup had little impact on tourism.[47] And like the controls placed on the movement of people across borders, restrictions on the movement of goods can be politicized in a way that generates significant international friction.

Even in the absence of trade restrictions, however, the economic burden that disease will place on developing countries will be severe. Added health care costs combined with a loss of productivity resulting from worker absences will exact a large economic toll. In 2001 the U.S. General Accounting Office (now the U.S. Government Accountability Office) estimated that Africa's total gross domestic product would now be one-third higher if malaria had been eradicated by 1970.[48]

The outbreak of disease can also lead a government to adopt policies that may be seen as discriminatory or politically motivated by segments of its own population—for example, treatment may be provided first, or exclusively, to a particular ethnic group, religious faction, or political party. This can provide antigovernment groups with the opportunity to increase their popularity and legitimacy by providing health services not provided by the government.[49] When these groups are sponsored by foreign governments, such as Iran's support for Hezbollah in Lebanon, the line between medicine and foreign policy vanishes.

In these economic and social circumstances, countries are susceptible to rapid political change. For instance, the inability or perceived unwillingness of political leaders to stop the spread of disease or to provide adequate care for the afflicted will undermine support for that government.[50] In countries with functioning democracies, this could lead to the election of new leaders with political agendas radically different from those of their predecessors. It could also breed greater support for populist candidates whose politics resonate in a society that believes that its economic and social hardships are due to neglect or mismanagement by the government. In countries with weak or nondemocratic political foundations, there is a heightened risk that this will lead to civil war or a toppling of the government altogether.

It is worth noting that Venezuela, a country with high geopolitical significance, could be hit hard by a climate-induced increase in disease as a result of increased rainfall that will create favorable conditions for many waterborne and vector-borne diseases. People living along Venezuela's coast, which will be subject to more frequent storms and flooding due to climate change, are at heightened risk.[51]

There is also a small chance that the balance of power between neighboring states could suddenly and decisively shift if one country's military or political elites were seriously affected by a disease while the other country's were not.[52] The high HIV infection rate in several African militaries, such as those of the Democratic Republic of the Congo, South Africa, and Angola, provides a recent example of how a disease can come to have a disproportionate impact on a sector of the population that is critical to a country's national security.[53]

Regardless of the scenario, however, developing countries will look to the United States and the developed world for help in responding to these health crises. The gap between the world's haves and have-nots will be made increasingly apparent, and the resentment that this will engender toward wealthy countries will only be assuaged if significant resources are devoted to combatting disease outbreaks and to caring for the afflicted in the developing world.

China's Climate Change Challenge

In the coming decades, climate change will pose a growing political and economic challenge to China. The Chinese leadership's response will have international security ramifications and will become a significant factor in determining the course of U.S.-Sino relations.

China's current pattern of energy production and consumption poses a tremendous long-term threat to the global environment. China is believed to have surpassed the United States as the world's largest national emitter of carbon dioxide—though, notably, it lags far behind on a per capita basis—and its energy demand is projected to grow at a rate several times that of the United States for decades to come (see table 4-2).

China's steep carbon emissions trajectory is to a large extent the result of its reliance on coal-fired power plants. Currently, coal constitutes approximately two-thirds of China's primary energy consumption, and it will continue to be a major fuel source for the foreseeable future because China has enormous coal reserves and coal is a far more cost-efficient energy source than imported natural gas at today's prices. China is now building traditional coal-fired power plants at a rate of almost one per week, each of which releases approximately 15,000 metric tons of CO_2 per day.[54]

Today, coal use accounts for more than 80 percent of China's carbon emissions, while automobile emissions account for only approximately 6 percent.[55] However, cars and trucks will be an increasingly important factor in the future: the size of China's vehicle fleet is projected to grow from 37 million to as many as 370 million during the next twenty-five years.[56]

Unless its pattern of energy consumption is altered, China's carbon emissions will reinforce or accelerate several existing domestic environmental challenges, ranging from desertification to water shortages to the deterioration of air quality in urban areas, and will also become the primary driver of global climate change itself. China's future will be shaped by how its leadership reacts to intensifying domestic and international pressure to deal with these challenges.

The IPCC projects in its 2007 *Fourth Assessment Report* that climate change will "impinge on sustainable development of most developing coun-

Table 4-2. Actual and Projected Energy Consumption and Carbon Emissions in China, United States, and World, 2003 and 2030

Selected energy use, 2003 actual/2030 projected, and percent annual growth rate

Metric	China	United States	World
Total energy (quadrillion BTUs)	45.5/139.1 4.2 %	98.1/166.2 1.3 %	420.7/721.6 2.0 %
Oil (million barrels/day)	5.6/15 3.8 %	20.1/27.6 1.2 %	80.1/118 1.4 %
Natural gas (trillion cubic feet)	1.2/7.0 6.8 %	22.3/26.9 0.7 %	95.5/182 2.4 %
Electricity (billion kilowatt hours)	1,671/5971 4.8 %	3,669/5619 1.6 %	14,781/30,116 2.6 %
Nuclear electricity (billion kilowatt hours)	42/304 7.6 %	764/871 0.5 %	2,523/3,299 1.0 %
Coal (million short tons)	2,168/5,855 3.7 %	1,095/1704 1.9 %	5,250/1,0581 2.5 %
CO_2 emissions (10^6 metric tons)	3,541/10,716 4.2 %	5,796/8,115 1.3 %	25,028/43,676 2.1 %
GDP per capita (dollars)	4,618/19,940 5.6 %	35,467/63,148 2.9 %	8,048/17,107 2.8 %
Per capita energy use (10^6BTUs per person)	31.5/96.2 4.2 %	337/454 1.1 %	66.7/88.0 1.0 %

Source: U.S. Energy Information Administration, *International Energy Outlook*, 2006.

tries of Asia, as it compounds the pressures on natural resources and the environment associated with rapid urbanisation, industrialisation, and economic development."[57] For instance, according to the report, "The rain-fed crops in the plains of North and North-East China could face water-related challenges in coming decades, due to increases in water demands and soil-moisture deficit associated with projected decline in precipitation."[58] China's first national report on climate change, released in late 2006, projected that national wheat, corn, and rice yields could decrease by as much as 37 percent in the next few decades.[59] Even a far smaller decrease, however, would require significant action by the central government.[60]

China, moreover, is severely affected by desertification, and the United Nations Framework Convention on Climate Change (UNFCCC) notes that desertification-prone countries are "particularly vulnerable to the adverse effects of climate change."[61] More than a quarter of China is already desert, and the Gobi is steadily expanding: it grew by about 52,400 square kilometers (20,436 square miles) between 1994 and 1999.[62] According to the United Nations Convention to Combat Desertification, this threatens the livelihoods of some 400 million people.[63]

Water shortages will also pose a major challenge to China. In 2004 the UN reported that most of China's major rivers had shrunk, and in December 2006 it found that the Yangtze River's water level dropped to an all-time low because of climate change.[64] Northern China faces the greatest threat in this respect, as it will be subject to heat waves and droughts that will worsen existing water shortages. In addition, two-thirds of China's cities are currently experiencing water shortages, and their predicament will be exacerbated by the shifts in precipitation patterns and increased water pollution from improperly treated industrial waste.[65]

In spite of the colossal development projects that China has initiated in an attempt to mitigate growing environmental stress—such as the South-to-North Water Diversion Project, which is anticipated to cost some $59 billion and take half a century to complete—domestic social and political turmoil will increase. One source of unrest will be increased human migration within China due to environmental factors. Much of this migration will reinforce the current migratory trends from countryside to city, putting added pressure on already overpopulated and dangerously polluted urban centers.[66]

Regions of China that benefit from some additional rainfall will also need to cope with an influx of migrants from water-scarce areas. In China's northwestern provinces, where rainfall may increase, the acceleration of the movement of Han Chinese into areas populated by the Uighur people, who are Muslim, will aggravate tensions that have led to low-level conflict for many years. This conflict has intensified as China has begun to extract natural resources from these provinces and as larger numbers of Han Chinese have migrated there in search of employment. The projected increase in Han migration to this area could provoke violent clashes and potentially lead to social turmoil.[67]

In the last few years, concerns over environmental issues have provoked thousands of Chinese to demonstrate across the country. In April 2005 as many as 60,000 people rioted in Huaxi Village in Zhejiang Province as a result of pollution from a chemical plant, and just three months later, 15,000 people rioted for three days in the eastern factory town of Xinchang—just 180 miles south of Shanghai—because of pollution from a pharmaceuticals factory.[68]

Moreover, the findings of a poll conducted in China last year by the Chicago Council on Global Affairs and WorldPublicOpinion.org indicate that there is widespread recognition among the Chinese public that climate change is a uniquely serious environmental problem. Some 80 percent of respondents concurred that within ten years, global warming could pose an important threat to their country's "vital interest."[69]

At present, robust economic growth is the bedrock of the Chinese leadership's domestic political strategy, but in the coming years the leadership will face growing public pressure to play a much more constructive role in managing the environment and dealing with its negative impacts. The Chinese people are likely to insist that their leaders assume greater responsibility for protecting the environment, addressing and redressing the economic damage that results from environmental degradation, and holding accountable those who violate environmental regulations.

On the one hand, this may lead to internal political reforms that will enable the country to more effectively cope with a range of environmental threats. To this end, the central government may assume a much larger role in affairs and policies that to date have been left largely in the hands of regional or local officials. At present, local officials of the Chinese State Environmental Protection Agency are selected not by high-level SEPA officials but by local governments.[70] These officials do not currently have the necessary incentive to enforce regulations that sacrifice short-term economic growth for longer-term environmental sustainability, and they are also vulnerable to corruption. If the government is to meet the underlying environmental challenges and enforce environmental regulations, it will need to change the incentive structure and provide more oversight.

However, it is also possible that the Chinese leadership will not make the necessary internal political reforms even as the effects of climate change and other environmental factors become increasingly severe. This could lead to larger protests and more frequent violent clashes with police, as well as more restrictions on the press and public use of the Internet. Relations with the West would rapidly deteriorate as a result.

A second factor that could shape China's future is not internal but external: the growing pressure from the international community to curb carbon emissions and to enter into a global carbon reduction agreement. To date, China has resisted policies and treaties that restrict its carbon emissions, opting instead to set its own targets for improved energy intensity (the ratio of energy consumption to GDP). The current national goal is to reduce energy intensity by 20 percent by 2010, and to quadruple GDP while only doubling energy use by 2020.[71] This target is considered extremely ambitious, and the added economic costs of constraining the country's carbon emissions would make it even more so.

In any event, there will be escalating pressure on China to be a "responsible stakeholder" as its economic and political strengths grow and its share of global carbon emissions increases. Mounting global awareness about the threats posed by climate change, as well as the harm it is inflicting on developing countries in which China is seeking to expand its political and economic influence, will make it difficult for China to remain outside a U.S.-supported post-Kyoto regulatory framework on climate change without severely damaging its international standing.

But if the United States is not a participant in the post-Kyoto framework and has not adopted significant carbon reduction policies of its own, China will undoubtedly be spared much of this international pressure and be far less likely to limit its carbon emissions, particularly given that climate change is just one of many environmental challenges it faces.

Impact of Climate Change on Fuel Types

In its 2006 International Energy Outlook, the U.S. Energy Information Administration (EIA) forecasts increased global demand for every major fuel type through 2030, though the rate of growth varies significantly among them (see figure 4-3).

This EIA projection provides a useful policy-neutral reference case for analyzing the pressures that climate change will exert on patterns of energy production and consumption. There will be significant foreign policy and national security implications for energy exporters and importers alike, including the following: a strengthened geopolitical hand for natural gas–exporting countries and, potentially, biofuel-exporting countries as well; a weakened hand, both strategically and economically, for importers of all fuel types, who will find themselves increasingly vulnerable to supply disruption; growing nuclear safety and proliferation threats; and a steady increase in the economic and environmental cost of delaying the implementation of global carbon reduction policies.

Oil

Climate change will exert upward pressure on oil prices by causing supply disruptions and contributing to instability in some oil-producing regions. The increase in temperature brought about by climate change will not result in a large enough reduction in the use of home heating oil to offset these effects because home heating oil constitutes a small percentage of global demand.[72]

Figure 4-3. World Marketed Energy Consumption by Region, 1980–2030

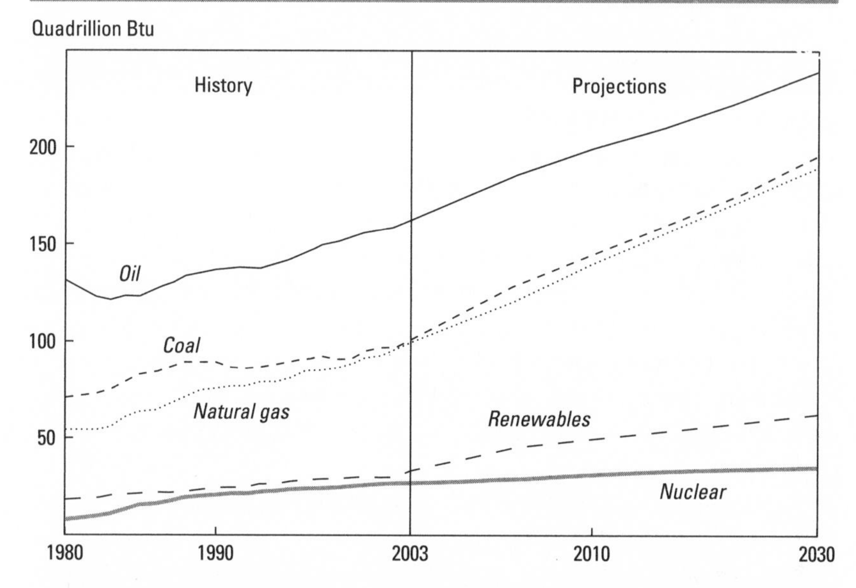

Source: U.S. Energy Information Administration, *International Energy Outlook*, 2006 (www.eia.doe.gov/oiaf/archive/ieo06/excel/figure_3data.xls).

In the IPCC's A1B greenhouse gas emission scenario, the increased frequency of major storms will lead to more damage to offshore rigs and coastal refineries, and oil tanker shipments will be delayed by weather events. Oil-exporting countries will benefit economically from the risk premium that climate change adds to the price of each barrel of oil.

Political instability in oil-exporting countries will be exacerbated by climate change as well, leading to reduced output due to factors ranging from acts of sabotage to lack of international investment. For instance, although the United States is currently projected to import between 25 and 40 percent of its oil from Africa by 2015, the adverse political and environmental conditions brought about by climate change may prevent Nigeria and the continent's nine other oil-exporting countries from expanding their existing oil production levels to meet this demand.[73]

Oil-importing developing countries, meanwhile, will be disproportionately affected by increases in the cost of oil because their economies have high-energy intensities—they require a great deal of energy per unit of GDP—and fuel switching is difficult. The International Energy Agency estimates that oil-importing and debt-burdened countries in sub-Saharan Africa will lose more than 3 percent of their GDP with each ten-dollar increase in the price of oil.[74]

In spite of rising oil prices and an expanding biofuels market, oil will remain a key strategic commodity for the United States, and the U.S. Navy will continue to protect global sea lanes in order to ensure the safe movement of oil shipments around the world. But as China develops its own blue-water navy in the next few decades it, too, will become involved in securing global sea lanes, in particular the routes linking Northeast and Southeast Asia.[75] As a result, the U.S. and Chinese navies will need to find ways of coordinating their movements if they are to avoid miscommunication or accidental interference that could cause severe political tension.[76]

U.S.-Sino relations could also be strained if China continues to supplement its international energy deals with state-to-state arrangements that include significant nonmarket elements such as building airports, offering credit, and tying foreign assistance to energy investment. To date the list of countries with which it has made such arrangements includes Angola, Sudan, Iran, Algeria, and Saudi Arabia.

A second growing concern for the United States is China's practice of investing in countries where sanctions and other factors limit or preclude the major Western international energy companies from operating. Although China may be motivated as much by economic as political factors—it is easier, after all, to compete in markets where there is less competition—such investment in sanctioned countries such as Sudan and Iran runs counter to the strategic interests of the United States. As China's demand for imported oil increases in the coming years, so will these investments.

Natural Gas

The upward pressure that climate change exerts on the price of oil is likely to help drive demand for natural gas. Furthermore, if stringent national or global carbon emission regulations are adopted, natural gas, as a less carbon-intensive energy source than coal or oil, will become an increasingly attractive fuel choice, particularly for electricity generation.

One likely development will be an increase in the size and scope of the liquefied natural gas (LNG) market. The United States' overseas LNG imports are poised to overtake imports from Canada as its primary source of imported natural gas within the next few years; Europe, China, and India have all been working to increase LNG imports as well.[77]

Although the development of an increasingly global LNG market will temper the strategic leverage of major natural gas exporters by providing some added security against targeted embargoes or price manipulation, the geopolitical power of countries that are rich in natural gas will nevertheless grow significantly by mid-century. This will create new security risks and new choke points around the world. Countries in Central Asia and the Caucasus will become more strategically important because they can offer energy supplies and routing alternatives to the Middle East and Russia.

It is Russia, however, that stands to benefit the most from the growing strategic significance of natural gas, as well as from the environmental impacts of climate change in general. Russia holds by far the world's largest proven natural gas reserves (almost twice those of Iran, the country with the second largest proven reserves) and currently supplies Europe with two-thirds of its imported natural gas.[78] A warmer climate will help to reduce domestic demand for energy: the IPCC anticipates that "in the United Kingdom and Russia a 2°C (3.6°F) warming by 2050 will decrease space heating needs in the winter, thus decreasing fossil fuel demand by 5–10 percent and electricity demand by 1–3 percent."[79] In the longer term, increased temperatures could also open up ice-locked northern shipping routes for the export of LNG and oil throughout the year.

During the past few years, Russia has proved willing to use its energy assets for political leverage. In January 2006, for instance, Russia dramatically increased the price of natural gas in the run-up to the Ukrainian parliamentary elections. Ukraine refused to pay the new rates, which led to a supply reduction that left it—and several EU countries that are supplied through pipelines that run through Ukraine's territory—short of natural gas in the middle of winter. As global demand for oil and natural gas grows, Russia's energy assets are likely to become an increasingly potent, and frequently employed, political tool.

This tension will be exacerbated (and become a more direct challenge to the national security of the United States) if NATO expands to include Ukraine, Georgia, or other countries that are embroiled in ongoing energy conflicts with Russia. Senator Richard Lugar (R-Ind.), who as chairman of the Senate Foreign Relations Committee did much to draw attention to global energy security threats, has argued that the deliberate cutoff of energy supplies to a NATO country should trigger a compulsory Article 5 collective response by its members.[80] According to this interpretation, Russia's natural gas–supply cutoff to Ukraine would have required U.S. action because Italy and other NATO allies were affected.

Another area of concern for the United States and its allies will be Russia's relationship with China. As Russia becomes an important supplier of energy to East Asia, the strategic interests of China and Russia may become more closely aligned, particularly with regard to Central Asia. Their joint leadership in the Shanghai Cooperation Organisation (SCO), a regional group whose membership also includes Kazakhstan, Kyrgyzstan, Tajikistan, and Uzbekistan, could enable them to exert significant influence over this critical region's energy supplies and pipelines as well as its overall political and strategic relationship with the West. For instance, at their July 2005 summit, SCO members issued a declaration calling for the closure of U.S. military bases in the region, and before the end of the month the United States had been formally evicted from its base in Uzbekistan.[81]

There also remains a more remote possibility that a natural gas cartel will develop out of the Gas Exporting Countries Forum, in which Russia plays a role analogous to that played by Saudi Arabia within OPEC. At present, natural gas is primarily distributed through pipelines that involve long-term regional contracts, and natural gas pricing is closely linked to oil prices. However, the International Energy Agency projects an expansion of global LNG capacity from 246 billion cubic meters a year in 2005 to 476 billion cubic meters by 2010. Simultaneously, a larger spot market for LNG will emerge, and this will make pricing more susceptible to manipulation by a cartel of natural gas suppliers. Russia, as the global natural gas giant, would stand to gain the most from the emergence of such a cartel.

Coal

For the first time in sixteen years of forecasting worldwide energy use, the 2006 International Energy Outlook projects that the rate of growth in coal consumption will exceed that of natural gas.[82] Although there is only one-tenth of a percent difference between their projected rates, this is an alarming development given the enormous environmental threat posed by carbon emissions from coal-fired power plants. In the absence of international carbon emission restraints, climate change will likely reinforce this trend by leading to an increase in the price of natural gas and oil relative to coal.

Coal's low cost as a fuel source for electricity generation and its wide distribution among developed and developing nations makes it inconceivable that coal can or will be largely replaced within the next thirty years.[83] Rather, the question is whether coal will continue to be a driver of climate change or whether the development and implementation of clean coal and, in particular, carbon capture and sequestration (CCS) technology can make it a viable fuel source in a carbon-constrained economy. A 2007 MIT study, "The Future of Coal," found that, in spite of the lead times involved, CCS technology can in fact be deployed on a wide enough scale to reduce significantly the carbon emissions from coal-fired power plants by 2050, though only if a global carbon emissions restriction or tax is in place and near-term government investment in R&D is increased.[84]

Nuclear Power

The EIA projects a slight decline in the installed nuclear capacity of OECD countries by 2030, but rapid growth in the nuclear sectors of non-OECD countries such as China.[85] Two of the factors that drive the use of nuclear power are high fossil fuel prices and energy insecurity. As we have seen, climate change will contribute to both.

There is a risk of proliferation associated with this fast expansion of nuclear power. The development of nuclear power capabilities and the associated facilities for the manufacturing and production of nuclear fuels could bring many more countries to the brink of possessing the capacity to develop nuclear weapons. There is also a smaller risk, that commercial fuel cycle technology will be transferred to a country that is interested in developing a clandestine nuclear weapons program.

Approximately a dozen countries in the Middle East and North Africa have recently sought the International Atomic Energy Agency's assistance in developing nuclear energy programs.[86] Political insecurity coupled with the increased availability of nuclear fuel cycle technology may lead these countries over time to pursue nuclear weapons programs as well.

There is also a risk that a Sunni Arab country will receive assistance from scientists or government officials from Pakistan, the only Sunni state that already possesses nuclear weapons. In addition, Bangladesh, a non-nuclear country, could be tempted to pursue such a program if climate change destabilizes the region and its relations with its nuclear neighbor, India, deteriorate further.

Furthermore, rapid nuclear expansion heightens the risk of a nuclear accident. In addition to the local health and environmental consequences, a large-scale accident anywhere in the world could provoke a global backlash against nuclear power. This would increase the economic burden of limiting carbon emissions by forcing countries to switch to more expensive alternatives and could cause countries to reconsider any carbon reduction policies in place.

If aggressive global carbon reduction policies are adopted in the coming years, nuclear energy will become more cost-competitive with fossil fuels. This could provide added political justification for countries to develop domestic commercial nuclear power programs that might lead to weapons programs or rekindle interest in weapons programs that had been abandoned. Despite these risks, however, nuclear power will continue to play an integral role in the energy strategies of many countries that are seeking to reduce their carbon emissions, making it all the more imperative that the international community redouble its nonproliferation efforts.

Biofuels

Under certain conditions, biofuels have the potential to emerge as a competitor to oil in the coming decades, particularly in the transportation sector. This is most likely to occur if a global carbon reduction policy is adopted that creates a strong market incentive for investments in both R&D and infrastructure for such fuels. The United States and Brazil currently account for more than 70 percent of global ethanol production, but other countries in Latin America and elsewhere could be poised to participate in an expanded international biofuels market.[87] This would help to offset some of the geostrategic importance of oil suppliers.[88] China could be a significant biomass fuel consumer, as it would rather import this fuel than sacrifice food crops for energy crops, particularly if its food security is threatened by climate change. Japan already imports ethanol from Brazil.

The biofuels market will need to be managed effectively in order for it to grow to scale and avoid replicating some of the flaws that plague the fossil fuel market. This requires developing and implementing policies that minimize the total "fields to wheels" carbon emissions from biofuels (which includes emissions from any fossil fuel used to raise energy crops, refine these crops into fuel, and distribute the fuel to consumers).[89] It is also important to consider nonenvironmental externalities, such as the impact that replacing food crops with energy crops could have on food prices around the world. Although to date productivity gains have enabled U.S. farmers to raise sufficient quantities of crops to meet demand for both food and fuel, policymakers will need to monitor this issue closely if demand continues to increase.

Disaster, Humanitarian, and Crisis Response: Challenges and Opportunities for the International Community

The natural disasters, humanitarian emergencies, and other crises that climate change creates or intensifies will present serious challenges not only to the directly affected countries but also to the entire international community.

The developing world will need substantial support to endure the effects of climate change, and it will seek this support with a full awareness that the historical responsibility for the high levels of anthropogenic carbon in the atmosphere rests on the developed world's shoulders.

The United Nations

As a result of climate change, the United Nations and other multinational organizations will be looked to with increased frequency to help manage refugee flows, food aid distribution, disaster relief, and other emergencies. To handle its increased workload, the UN will need increased financial and diplomatic support. The United States is likely to supply the former consistently but the latter inconsistently, as operations that require the consent of the UN Security Council will invariably become entangled in disparate international political disputes.

The UN will also be called upon to play a central role in negotiating and implementing a post-Kyoto international carbon emissions reduction scheme. As the effects of climate change become more serious and disruptive, calls for unified global action will grow ever louder, and a failure to reach a meaningful consensus could precipitate wider political breakdowns at the world body.

For instance, when the UN Security Council in April 2007 decided to take up the issue of climate change and energy security, 135 developing countries quickly united to protest what they saw as a hypocritical effort by some of the world's worst emitters, past and present, to wrest control of the climate change issue from the General Assembly.[90] This could foreshadow much more acrimonious clashes between large and small emitters and even between developed nations with differing emissions policies.

In the future, the UN might seek to avert these clashes between the Security Council and the General Assembly by creating a new "Climate Security Council" in which key developed and developing countries such as Germany, Japan, Brazil, India, China, and South Africa would be represented. If the UN fails to provide an effective institutional setting for debate and decision making on climate change issues, there will be increased interest in developing alternative forums. For example, an "E-8" forum may emerge whose membership would be the world's major carbon emitters and which would be devoted exclusively to ecological and resource issues.[91]

In addition to struggles within and between the Security Council and General Assembly, a serious challenge to the UN is likely to be the magnitude of the demands placed upon it by environmental migration. In the aftermath of World War II, the UN established a system to protect civilians who had been forced from their home countries by political violence. Today, there are almost 9 million officially designated refugees under the protection of the Office of the UN High Commissioner for Refugees (UNHCR), but this number is dwarfed by the more than 25 million people who have fled their homes as a result of environmental degradation.[92] The IPCC estimates that the number could reach 50 million by the end of the decade, and up to 200 million by 2050, though even a far smaller figure will prove difficult for the UNHCR to manage.[93]

The UNHCR has thus far refused to grant these people refugee status, instead designating them "environmental migrants," in large part because it lacks the resources to meet their needs. But with no organized effort to supervise the migrant population, these desperate individuals go where they can, not necessarily where they should. As their numbers grow, it will become increasingly difficult for the international community to ignore this challenge. Significantly more resources will need to be channeled to the UNHCR as well as to other critical international bodies, in particular the International Red Cross and Red Crescent organizations.

The European Union

The European Union today is at the forefront of action to reduce the greenhouse gas emissions of major economies. Its member states continue to lead the international community in carbon reduction policies and practices. The entire EU is responsible for only 14 percent of global carbon emissions at present, and this percentage will shrink even further in coming years.[94] It has also established the world's first functioning carbon market, which could evolve into a global one in years to come. Already the EU is considering expanding its Emissions Trading Scheme to include subnational states or entities such as the state of California.

Consequently, it is likely that the EU will cement its position as the most responsible and united regional organization on the issue of climate change. The Organization of American States may rival the EU in terms of its member countries' carbon emissions, but the OAS is not structured to formulate and enforce institution-wide policies, and that seems unlikely to change. Likewise, though the ASEAN Regional Forum (which is made up of twenty-seven countries that have a bearing on the security of the Asia Pacific region) and the East Asia Summit (in which both China and India are participants) bring many of the world's worst carbon emitters together to cooperate on energy and economic issues, these organizations lack the capacity and the mandate needed to develop and impose carbon reduction policies on their members.

United States as "First Responder"

Although some of the emergencies created or worsened by climate change may ultimately be managed by the UN, the United States will often be looked to as a "first responder" in the immediate aftermath of a major natural disaster or humanitarian emergency. The larger and more logistically difficult the operation, the more urgent the appeal will be.

The question of whether and how to respond will be a recurring one for the United States, each time raising a difficult set of questions with important national security and foreign policy implications: How much financial assistance should the United States pledge and how quickly? With which other countries should the United States seek to coordinate its response, either operationally or diplomatically? Should the U.S. military participate directly, and if so, in what capacity and on what scale?

This last question is particularly sensitive, but it presents potential geopolitical rewards as well as risks. For instance, the U.S. military played a vital role in the international relief efforts undertaken in the aftermath of the December 2004 Indian Ocean tsunami. There was simply no substitute for the more than 15,000 U.S. troops, two dozen U.S. ships, and 100 U.S. aircraft that carried out the operation.

The performance of the U.S. military was resoundingly applauded by the international community. In Indonesia itself, the public image of the United States improved dramatically: a Pew Research Center poll conducted in the spring of 2005 found that 79 percent of Indonesians had a more favorable impression of the United States because of its disaster relief efforts, and as a result the United States' overall favorability rating in Indonesia rose to 38 percent after having bottomed out at 15 percent in May 2003.[95] U.S. Admiral Michael Mullen, chairman of the Joint Chiefs of Staff, was right to describe the military's response to the tsunami and the subsequent improvement of America's image in the region as "one of the most defining moments of this new century."[96]

But it is not yet clear whether the tsunami response will be remembered in thirty years' time as "defining" or as an exceptional case. As the world looks to the United States for assistance with greater frequency, and when disaster strikes in places where the U.S. military could be greeted with some hostility, executing relief missions will become increasingly complex and dangerous.

What will happen when a U.S. soldier or marine is killed by an insurgent or terrorist in the midst of a relief operation? Will the United States shun direct participation in countries where it fears that short-term humanitarian assistance could evolve into long-term stability operations, even if it is precisely these countries that are in the greatest danger of failing without such direct engagement?

It is not only shifting international and domestic political circumstances that will present new challenges to the U.S. military: the shifting physical environment will do so as well. The increased frequency of severe storms will create adverse conditions, particularly for air and sea operations, while rising sea levels may threaten the long-term viability of bases situated on islands or low-lying coastal areas.

Consequently, the U.S. military will need to plan for how it would protect or, in extreme circumstances, compensate for the loss of bases in vital strategic areas such as the Diego Garcia atoll in the southern Indian Ocean, which serves as a major hub for U.S. and British missions in the Middle East and was instrumental in the military's rapid response to the tsunami.[97] Expanding existing bases or establishing new ones can be both expensive and politically treacherous, and it is possible that the United States will choose to invest more in developing its own offshore "sea basing" platforms that do not require host country consent.

The roles of the U.S. Army and National Guard will also need to evolve. At present, National Guard troops are responsible for responding to domestic natural disasters when needed, yet their deployment overseas (for either military or relief operations) can leave the United States short of troops and equipment precisely when climate change will be causing more extreme weather events domestically. Furthermore, regular Army and Marine Corps troops may need to receive training in how to provide disaster relief in potentially hostile environments, perhaps as part of a post-Iraq focus on developing the skill sets needed for counterinsurgency, stabilization, and other nonconventional operations.

More generally, it is possible that the United States will become reluctant to expend ever greater resources on overseas disaster relief, not to mention longer-term humanitarian and stabilization operations, as the impact of climate change at home becomes increasingly acute. Natural disasters already cost the United States billions of dollars annually, and the IPCC projects that climate change will create an "extended period of high fire risk and large increases in area burned" in North America and particularly in the western United States.[98] In addition, the United States will have to meet rising health costs associated with more frequent heat waves, a deterioration of air quality, and an increase in waterborne disease.

The Danger of Desensitization

In the course of the next three decades, the spread and advancement of information and communication technologies will enable people to follow international crises ever more closely, making it increasingly difficult to ignore the disparity between how the world's haves and have-nots are affected by climate change. As noted in a recent report by the UK Ministry of Defense's Development, Concepts, and Doctrine Center, however, the very words and images that at first will catalyze action could over time lose their impact: "Societies in the developed and developing worlds *may* become increasingly inured to stories of conflict, famine and death in these areas and, to an extent, desensitized."[99]

Ultimately, the threat of desensitization could prove one of the gravest threats of all, for it is clear that the national security and foreign policy challenges posed by climate change are tightly interwoven with the moral challenge of helping those least responsible to cope with its effects. And if the international community fails to meet either set of challenges, it will fail to meet them both.

Conclusion

The effects of climate change we describe in this scenario are not alarmist; rather, they are to a large degree inescapable. The scientific evidence is clear that we will see environmental shifts at least as dramatic as those we outline here. What is not inevitable, however, is how human society responds to global warming and its attendant resource scarcity, extreme weather, and increase in the incidence of disease. It is critical that governments, particularly in the wealthier nations that have the requisite tools and resources, begin to plan on an urgent basis for ways to prevent, mitigate, and manage the risks of climate change that we have outlined here. Delaying this planning process risks touching off a chain reaction of crisis that will be extremely difficult to stop once it is firmly under way.

Notes

1. This chapter was prepared with research assistance from Lillian Smith.
2. Idean Salehyan, "Refugees, Climate Change, and Instability," Human Security and Climate Change International Workshop (Oslo: 2005), pp. 1–10.

3. As of January 2008 the federal government had spent in excess of $100 billion in efforts to repair the damage caused by Hurricane Katrina. For more information, see the regularly updated "New Orleans Index" issued by the Brookings Institution (http://www.gnocdc.org/NOLAIndex/NOLAIndex.pdf).

4. Celine Charveriat, "Natural Disasters in Latin America and the Caribbean: An Overview of Risk," Working Paper 434 (Washington: Inter-American Development Bank, 2000), p. 58.

5. Salehyan, "Refugees, Climate Change, and Instability," p. 15.

6. Ibid., pp. 4–6, 10.

7. IPCC, "Summary for Policymakers," in *Climate Change 2007: Impacts, Adaptation and Vulnerability: Contribution of Working Group II to the Fourth Assessment Report of the Intergovernmental Panel on Climate Change,* edited by M. L. Parry and others (Cambridge University Press, 2007), p. 13.

8. Tom Felix Joehnk, "The Great Wall of India," *Economist,* special report, "The World in 2007," March 2007, p. 49; Jon Barnett, "Security and Climate Change," Working Paper 7 (Norwich, England: Tyndall Centre for Climate Change Research, 2001), pp. 4–5.

9. Chicago Council on Global Affairs and WorldPublicOpinion.org, "Poll Finds Worldwide Agreement That Climate Change Is a Threat: Publics Divide over Whether Costly Steps Are Needed" (Chicago: 2007) (www.worldpublicopinion.org/pipa/pdf/mar07/CCGA+_ClimateChange_article.pdf).

10. Ibid.

11. R. V. Cruz and others, "Asia," in *Climate Change 2007: Impacts, Adaptation and Vulnerability,* p. 472.

12. Ibid., p. 484.

13. Somini Sengupta, "Sea's Rise in India Buries Island and a Way of Life," *New York Times,* April 11, 2007 (www.nytimes.com/2007/04/11/world/asia/11india.html?_r=1&oref=slogin).

14. Joehnk, "Great Wall of India," p. 49.

15. Sudha Ramachandran, "The Threat of Islamic Extremism to Bangladesh," *Power and Interest News Reporting,* July 27, 2005 (www.pinr.com/report.php?ac=view_report&report_id=334&language_id=1aol_htm).

16. Bruce Riedel, "Al Qaeda Strikes Back," *Foreign Affairs,* May–June 2007. Riedel argues that Bangladesh is among the places most likely to become a new base of operations for al Qaeda (see pp. 24–70).

17. World Bank, "Clean Energy and Development: Towards an Investment Framework, Annex K," report prepared by World Bank staff for the Development Committee, April 5, 2006 (http://siteresources.worldbank.org/DEVCOMMINT/Documentation/20890696/DC2006-0002(E)-CleanEnergy.pdf).

18. Michael McCarthy, "Climate Change Will Cause 'Refugee Crisis,' " *Independent,* October 20, 2006 (www.independent.co.uk/environment/climate-change/climate-change-will-cause-refugee-crisis-420845.html).

19. M. Boko and others, "Africa," in *Climate Change 2007: Impacts, Adaptation and Vulnerability,* pp. 433–67.

20. Elizabeth Leahy, ed., *The Shape of Things to Come: Why Age Structure Matters to a Safer, More Equitable World* (Washington: Population Action International, 2007).

21. Jad Mouawad, "Growing Unrest Posing a Threat to Nigerian Oil," *New York Times,* April 21, 2007 (www.nytimes.com/2007/04/21/business/worldbusiness/21oil.html).

22. Ibid.

23. Michael Case, "Climate Change Impacts on East Africa," World Wildlife Fund Issue Brief, November 2006, p. 4.

24. Ibid.

25. CNA Corporation, *National Security and the Threat of Climate Change* (Alexandria, Va.: 2007), p. 15 (http://securityandclimate.cna.org/report/National%20Security%20and%20the%20Threat%20of%20Climate%20Change.pdf).

26. IPCC, "Summary for Policymakers," in *Climate Change 2007: Impacts, Adaptation and Vulnerability,* p. 8.

27. Robert Leiken, "Europe's Angry Muslims," *Foreign Affairs* 84 no. 4 (July–August 2005): 120–35.

28. Ibid., p. 1.

29. AP, "EU Ministers Gather for Debate on African Migration," *Taipie Times,* September 30, 2006, p. 6.

30. "Spain Vows to Curb Migrant Wave," BBC News, September 4, 2006.

31. "EU Unveils News Immigration Plans," BBC News, November 30, 2006.

32. Amy Otchet, "Saber-Rattling among Thirsty Nations," *UNESCO Courier,* October 2001 (www.unesco.org/courier/2001_10/uk/doss01.htm).

33. Michael Schultz, "Turkey, Syria, and Iraq: A Hyrdopolitical Security Complex," in *Hydropolitics: Conflicts over Water as a Development Constraint,* edited by L. Ohlsson (Atlantic Highlands, N.J.: Zed Books, 1995): 107–13.

34. Farzaneh Roudi-Fahimi and others, "Finding the Balance: Population and Water Scarcity in the Middle East and North Africa," report (Washington: Population Reference Bureau, 2002), pp. 1–2.

35. Alex Handcock, "Water Conflict: A Critical Analysis of the Role of Water in the Middle East," 2004, pp. 1, 9 (www.amcips.org/PDF_books/BookIV19.pdf).

36. Turkey demonstrated its readiness to cut water supply to Syria in January 1990, when it disrupted the flow of the Euphrates River in order to fill a reservoir in front of the Ataturk Dam; Handcock, "Water Conflict."

37. Maher Bitar, "Water and the Palestinian-Israeli Conflict: Competition or Cooperation?" (Washington: Foundation for Middle East Peace, December 22, 2005).

38. Ayca Ariyoruk, "Turkish Water to Israel?" (Washington: Washington Institute for Near East Policy, August 14, 2003). Israel has determined that for the time being

it can meet its water needs more efficiently by boosting its domestic desalinization capacity.

39. Israel, Jordan, and the Palestinian Authority have also agreed to conduct a feasibility study of a "Two Seas Canal" that would move water from the Red Sea to the Dead Sea. However, there are significant concerns that the project would pose a threat to existing aquifers beneath the new canal. See Mati Milstein, "Diverting Red Sea to Save Dead Sea Could Create Environmental Crisis," *National Geographic News*, December 14, 2006 (http://news.nationalgeographic.com/news/2006/12/061214-dead-sea.html).

40. Roudi-Fahimi and others, "Finding the Balance, pp. 2–3.

41. Jan Selby, "The Geopolitics of Water in the Middle East: Fantasies and Realities," *Third World Quarterly* 26 (2005): 329, 343–44 (www.sussex.ac.uk/Users/js208/thirdworldquarterly.pdf).

42. Andrew Jack, "Climate Toll 'to Double within 25 Years,' " *Financial Times*, April 24, 2007 (www.ft.com/cms/s/0/054ff7ca-f282-11db-a454-000b5df10621.html).

43. Assaf Anyamba and others, "Developing Global Climate Anomalies Suggest Potential Disease Risks for 2006–2007," *International Journal of Health Geographics* 5 (2006): 5.

44. Ibid. Conversely, some airborne diseases will thrive in precisely those areas that become more arid as a result of drought and higher temperatures, such as in parts of Brazil.

45. In 1834 France proposed a meeting with the other major European powers in order to standardize quarantining practices. The First International Sanitary Conference was convened in 1851, and after ten subsequent conferences the European powers eventually reached a consensus on minimum and maximum detention periods, as well as on international disease notification procedures. See Krista Maglen, "Politics of Quarantine in the 19th Century," *Journal of the American Medical Association* 290 (2003): 2873.

46. Gary Cecchine and Melinda Moore, "Infectious Disease and National Security: Strategic Information Needs," report prepared for the Office of the Secretary of Defense (Santa Monica: Rand Corporation, 2006), pp. 19–21.

47. "SARS, Thailand, Tourism and Business Travel: How Fast for Recovery?" *Asian Market Research News*, April 30, 2003.

48. U.S. General Accounting Office, "Global Health: Challenges in Improving Infectious Disease Surveillance Systems" (Washington: 2001) (www.gao.gov/new.items/d01722.pdf).

49. The increase in the number of orphaned or abandoned children that can be a further result of disease outbreaks may also pose a long-term security challenge, as these parentless children may be more susceptible to radicalization. See Susan Peterson, "Epidemic Disease and National Security," *Security Studies* 12, no. 2 (Winter 2002–3): 43, 60–61.

50. Cecchine and Moore, "Infectious Disease and National Security," pp. 17–18.

51. Andrew Simms and Hannah Reid, *Up in Smoke? Latin America and the Caribbean: The Threat from Climate Change to the Environment and Human Development* (London: International Institute for Environment and Development, Working Group on Climate Change and Development, 2006), pp. 5, 18, 39.

52. Peterson, "Epidemic Disease and National Security," p. 55.

53. Stefan Lovgren, "African Army Hastening HIV/AIDS Spread," *Jenda: A Journal of Culture and African Women Studies* (2001): 1, 2 (www.jendajournal.com/vol1.2/lovgren.html).

54. John Podesta, John Deutch, and Peter Ogden, "China's Energy Challenge," in *China's March on the 21st Century,* report, edited by Kurt Campbell and Willow Darsie (Washington: Aspen Strategy Group, June 1, 2007), p. 55.

55. Daniel Rosen and Trevor Houser, "China Energy: A Guide for the Perplexed," *China Balance Sheet,* May 2007, p. 15.

56. Ibid., p. 34.

57. Cruz and others, "Asia," p. 471.

58. Ibid., p. 482.

59. Ling Li, "China Releases First National Report on Climate Change," World Watch Institute (Washington: January 11, 2007) (www.worldwatch.org/node/4848).

60. In this climate change scenario (unlike in more severe scenarios), there is less risk that the impact of climate change on the ocean will generate a significant strain on the availability of ocean-sourced food in the near future, and thus the likelihood of "protein wars" is low. However, as some countries begin to restrict fishing in their coastal waters to prevent depopulation, international competition among fishermen will begin to grow. For example, European fishermen have responded to local restrictions by increasing their presence off the coast of West Africa, which has put them into more direct competition with Chinese fishermen in that region. See Lester Brown, "Fisheries Collapsing," *Eco-Economy: Building an Economy for the Earth* (New York: Norton, 2001).

61. United Nations, "United Nations Framework Convention on Climate Change" (New York: 1992) (http://unfccc.int/resource/docs/convkp/conveng.pdf), p. 2.

62. United Nations Environment Program, "The Environment in the News" (New York: 2004) (www.unep.org/cpi/briefs/Brief01April04.doc), p. 25.

63. United Nations, "China National Report on the Implementation of United Nations Convention to Combat Desertification and National Action Programme to Combat Desertification" (New York: 2000) (www.unccd.int/cop/reports/asia/national/2000/china-summary-eng.pdf), p. 1.

64. Peter Harmsen, "Dire Warnings from First Chinese Climate Change Report," Agence France-Presse, December 27, 2006.

65. Jonathan Lewis, producer, "Shifting Nature: China's Environmental Future," episode 4 of *China from the Inside,* series prepared for KQED and Granada Television (www.pbs.org/kqed/chinainside/nature/environment.html).

66. Climate change will not be a key determinant of the level of migration from North Korea to China. Although climate change could adversely affect food production in North Korea, domestic political and economic conditions will be the decisive factors in any future migration.

67. Inventory of Conflict and Environment, "Ethnic Conflict and Natural Resources: Xinjiang, China," ICE case study 183, May 2006 (www.american.edu/ted/ice/xinjiang.htm).

68. Howard French, "Anger in China Rises over Threat to Environment," *New York Times*, July 19, 2005; Jim Yardley, "Thousands of Chinese Villagers Protest Factory Pollution," *New York Times*, April 13, 2005 (www.nytimes.com/2005/04/13/international/asia/13cnd-riot.html).

69. Chicago Council on Global Affairs and WorldPublicOpinion.org, "Poll Finds Worldwide Agreement That Climate Change Is a Threat," March 13, 2007: "Eighty-three percent [of the respondents] say steps should be taken to address global warming. Of these, 42 percent believe it is a 'serious and pressing problem' that demands immediate action 'even if this involves significant costs' and 41 percent say the effects will be gradual and should be dealt with through 'steps that are low in cost.' Most Chinese (80%) think global warming could be an important threat to their country's 'vital interests' in the next ten years. Nearly half consider it a 'critical threat.' . . . Four in five Chinese respondents (79%) say that if developed countries are willing to provide aid, 'less-developed countries should make a commitment to limit greenhouse gas emissions."

70. Jared Diamond, *Collapse: How Societies Choose to Fail or Succeed* (New York: Viking, 2005), p. 375. Diamond also observes that the government could change the calculation by increasing the price of key environmental resources. At present, prices are set "so low as to encourage waste: e.g., a ton of Yellow River water for use in irrigation costs only between 1/10 and 1/100 of small bottle of spring water, thereby removing any financial incentive for farmers to conserve water" (p. 375).

71. Fredrich Kahrl and David Roland-Holst, "China's Carbon Challenge: Insights from the Electric Power Sector," Research Paper 110106 (University of California–Berkeley, Center for Energy, Resources, and Economic Stability, November 2006), pp. 17, 33 (http//are.berkeley.edu/~dwrh/CERES_Web/Docs/CCC_110106.pdf).

72. In the United States, home heating oil accounts for less than 1 million barrels per day of oil consumption. See Energy Information Administration, "Annual Energy Review 2004" (Washington: August 2005) (www.grinzo.com/energy_old/downloads/us_flow_diagrams_2004.pdf).

73. CNA Corporation, *National Security and the Threat of Climate Change*, p. 20.

74. Kathy Roche and Teresita Perez, *Our Addiction to Oil Is Fueling World Poverty* (Washington: Center for American Progress, April 6, 2006).

75. "US Navy Commander in Japan Says China Will Safeguard Global Sea Lanes," Associated Press, May 3, 2007.

76. Ibid.

77. Energy Information Administration, *International Energy Outlook 2006* (Washington: 2006) (www.fypower.org/pdf/EIA_IntlEnergyOutlook(2006).pdf), pp. 39–41.

78. Energy Information Administration, "Country Analysis Brief: Russia" (www.eia.doe.gov/emeu/cabs/Russia/NaturalGas.html).

79. J. Alcamo and others, "Europe," in *Climate Change 2007: Impacts, Adaptation and Vulnerability,* p. 556.

80. Richard Lugar, "Speech in Advance of NATO Summit," November 22, 2006 (http://lugar.senate.gov/pressapp/record.cfm?id=266087).

81. Robin Wright and Ann Scott Tyson, "U.S. Evicted from Air Base in Uzbekistan," *Washington Post,* July 30, 2005, p. A1.

82. Energy Information Administration, *International Energy Outlook 2006* (Washington: 2006), p. 3.

83. "The Future of Coal," report prepared by a study group at the Massachusetts Institute of Technology (Cambridge, Mass.: 2007), p. 7.

84. Ibid. According to "The Future of Coal," this process cannot occur overnight. Even in the most favorable circumstances, "Many years of development and demonstration will be required to prepare for [the] successful, large scale adoption [of the necessary coal technology]" (15).

85. Energy Information Administration, *International Energy Outlook 2006,* pp. 68–69.

86. William J. Broad and David E. Sanger, "With Eye on Iran, Rivals Also Want Nuclear Power," *New York Times,* April 15, 2007.

87. As we have seen in the United States in the past five years, farmers can switch rapidly from food crops to energy crops when the proper market incentives are present.

88. (S&T)2 Consultants Inc., "Issue Paper on Biofuels in Latin America and the Caribbean," paper prepared for the Inter-American Development Bank (Washington: September 2006).

89. David Tilman and Jason Hill, "Corn Can't Solve Our Problem," *Washington Post,* March 25, 2007, p. B1.

90. Thalif Deen, "Security Council Accused of Overstepping Bounds," Inter Press Service News Agency, April 12, 2007 (http://ipsnews.net/news.asp?idnews=37334).

91. For information on the potential role of the E-8, see Todd Stern and William Antholis, "Action Memorandum: Creating the E8," *American Interest,* January 2007 (www.brookings.edu/views/articles/antholis/200701.htm).

92. Office of the UN High Commissioner on Refugees, "The State of the World's Refugees 2006" (New York: 2005) (www.unhcr.org/publ/PUBL/4444d3bf25.html); BBC News, "Millions 'Will Flee Degradation,' " October 11, 2005.

93. Jonathan Leake, "Climate Change 'Could Create 200M Refugees,' " *Sunday Times (*London), April 1, 2007 (www.timesonline.co.uk/tol/news/uk/science/article 1596769.ece).

94. Ian Traynor and David Gow, "EU Promises 20% Reduction in Carbon Emissions by 2020," *Guardian,* February 21, 2007 (http://environment.guardian.co.uk/climatechange/story/0,,2017600,00.html).

95. Pew Global Attitudes Project, "American Character Gets Mixed Reviews," June 2005 (http://pewglobal.org/reports/display.php?PageID=801).

96. Mike Mullen, "What I believe: Eight Tenets That Guide My Vision for 21st Century Navy," *Proceedings,* January 2006 (www.navy.mil/navydata/cno/mullen/proceedingsjan06.html).

97. CNA Corporation, *National Security and the Threat of Climate Change,* p. 48.

98. IPCC, "Summary for Policymakers," in *Climate Change 2007: Impacts, Adaptation and Vulnerability,* p. 14; Timothy Gardner, "Warming Could Spark North American Water Scramble, UN Warns," Reuters, April 12, 2007 (www.enn.com/globe.html?id=1576).

99. UK Ministry of Defence, Development, Concepts, and Doctrine Center, "The DCDC Global Strategic Trends Programme" (London: 2007), p. 41.

five

SECURITY IMPLICATIONS OF CLIMATE SCENARIO 2
Severe Climate Change over the Next Thirty Years

LEON FUERTH

Scenarios are exercises of the imagination, whether the objective is to produce a movie or run a war game. What follows is a double scenario. At its core is a visualization of the physical consequences that might ensue from a particular level of global climate change. These consequences have a certain credibility because they are based on cutting-edge scientific analysis. Wrapped around this core, however, is my effort to imagine what political, economic, and societal consequences might be set in motion by these physical events. In this case, there can be no appeal to science: the reader is only called upon to "suspend disbelief" long enough to think about where we may all be heading. That we are indeed moving toward the unknown is a fact that demands our urgent attention.

Scenario Overview: Severe Climate Change

The recently released Intergovernmental Panel on Climate Change (IPCC) studies are a substantial improvement over their predecessors: they now assign probabilities to alternative physical outcomes of climate change, and they express these alternatives within a time frame. That said, it is important to keep in mind that the IPCC models still provide only a starting point for estimating the national security implications of climate change.

Much more needs to be done before it will be possible to convert broad-gauged forecasts of climate change into relatively precise forecasts of local climatic changes, from which it would then be possible to assess the economic, social, and political consequences. Nevertheless, climate change itself is no longer debatable and prospective: it is under way, and every week brings word that rates of change are much faster than had been anticipated.

The IPCC reports make it possible to chart at least a notional timeline for significant environmental and ecological change over the next century. There are considerable uncertainties as to the timing and magnitude of events along this line, but we have enough data to warrant making at least approximate correlations between these events and their policy and security implications.

The projection of severe climate change employed in this chapter is based on IPCC findings,[1] with an adjustment to account for possible "tipping point" events such as the abrupt release of massive quantities of methane from melting tundra or of carbon dioxide as the sea warms up. Under these conditions, adverse trends could accelerate abruptly, as follows:

—Over the next thirty years, average global surface temperature rises unexpectedly to 2.6°C (4.7°F) above 1990 levels, with larger warming over land and at high latitudes. Dynamic changes in polar ice sheets accelerate rapidly, resulting in about 52 centimeters (20 inches) of sea level rise. On the basis of these observations and of improved understanding of ice sheet dynamics, climate scientists express high confidence that the Greenland and West Antarctic ice sheets have been destabilized and that 4 to 6 meters (13 to 20 feet) of sea level rise are now inevitable over the next few centuries, bringing intense international focus to this problem.

—Water availability decreases strongly in the most affected regions at lower latitudes (dry tropics and subtropics), affecting 1 billion to 2 billion people worldwide. The North Atlantic meridional overturning circulation (MOC) slows significantly, with consequences for marine ecosystem productivity and fisheries.

—Crop yields decline significantly in the fertile river deltas because of sea level rise and damage from increased storm surges. Agriculture becomes essentially nonviable in the dry subtropics, where irrigation becomes exceptionally difficult because of dwindling water supplies, and soil salinization is exacerbated by more rapid evaporation of water from irrigated fields. Arid regions in the low latitudes have spread significantly by desertification, taking previously marginally productive crop lands out of production.

—Global fisheries are affected by widespread coral bleaching, ocean acidification, substantial loss of coastal nursery wetlands, and warming and drying of tributaries that serve as breeding grounds for anadromous fish (ocean fish that breed in freshwater streams).

—The Arctic Ocean is now navigable for much of the year because of decreased Arctic sea ice, and the Arctic marine ecosystem is dramatically altered. Developing nations at lower latitudes are impacted most severely because of climate sensitivity and high vulnerability. Industrialized nations to the north experience net harm from warming and must expend greater proportions of GDP adapting to climate change at home.

This projection serves as the basis for a scenario depicting the possible societal consequences of severe climate change over the course of thirty years. These consequences are not to be taken as predictions: they represent a selected construct of the future, intended to encourage reflection about the consequences of continued inaction.

The Role of Complexity

Climate change is a manifestation of phenomena that are complex in the technical sense of that word. Complex phenomena are nonlinear and unstable. *Nonlinear* means that incremental change in the level of inputs to a system can result in major, and even discontinuous, changes in the system's output. *Unstable* means that it is not possible to create a single, normative model for the system's behavior: instead, modeling must assume the possibility of surprise. It is readily seen that even incremental levels of climate change will have political consequences, but a less obvious, and major, premise of this chapter is that *nonlinear climate change will produce nonlinear political events.*

All structures exist within an operating environment. If that environment changes beyond some critical point, the structures that are adapted to it will break down. This applies not only to the physical world but also to social organization. Our systems of governance, including everything from individual nation-states to multinational systems, all operate within tolerances. Climate is a dimension of those tolerances.

One should not rule out the possibility that the cumulative "sociopolitical" effect of climate shock will be for the good, galvanizing massive efforts to change course through new technology and new growth strategies. There are ways to bias outcomes in this direction, but these have not yet been put in place, and the time for doing so is fleeting. Therefore, my assumption is that the political response will be biased toward outcomes that reflect pessimism and even panic.

As noted earlier, if the environment deteriorates beyond some critical point, natural systems that are adapted to it will break down. This applies also to social organization. Beyond a certain level climate change becomes a profound challenge to the foundations of the global industrial civilization that is the mark of our species.

Regional Sensitivity to Severe Climate Change

According to the IPCC findings, the poorest nations will suffer first and also most deeply from climate change. Despite this, my analysis of the international consequences of climate change begins with the wealthiest and strongest societies, since it is their responses that will make the difference between relative order and freefall.

United States

Even at lesser degrees of climate change we should expect more severe weather along our coasts, with increasingly violent storms coming in from the sea at much higher rates of incidence. Very early on in this process important social readjustments will occur—if only because of measures that the insurance and mortgage industries will take in their own defense. This is already visible along the Gulf Coast in Hurricane Katrina's aftermath.

Even at linear rates of sea level rise, such as those forecast at the lower range of the scenario, exponentially greater numbers of people would be affected. One storm model concludes that what is now a hundred-year flooding event in New York City will, with an additional 1 meter (3.3 feet) of sea level, be a four-year event.[2] Early on, there will be talk of massive engineering efforts to protect major economic centers along the coasts, including oil and gas production in the Gulf. In our scenario, however, estimates of conditions abruptly become worse as science adjusts for new theory and new data. Given this deteriorating prospect for the future, the idea of resisting nature by brute engineering will give way to strategic withdrawal, combined with a rear guard action to protect the most valuable of our assets. Optimists might hope for a gradual relocation of investment and settlement from increasingly vulnerable coastal areas. After a certain point, however, sudden depopulation may occur.

Severe climate change will attack the West Coast's economic foundations because of drastic, permanent water shortage, resulting not only from reduced annual rainfall but also from the disappearance of mountain snow, whose spring melt-off is vital to the entire region's hydrology. The water requirements of the great West Coast cities are already in conflict with the requirements of the region's agriculture. In the more destructive ranges of the severe scenario, it would no longer be possible to bridge this conflict through political compromise or adroit water management. Political tensions would be severe. Moreover, the damage to American agriculture will not be limited to California. There will be intensified dependence on irrigated farming in the Midwest, and this will result in the accelerated depletion of the Ogallala Aquifer, upon which the entire region's agrarian economies depend.[3]

The United States' federal system may also experience stress. As noted above, one possible consequence of severe climate change will be greatly increased frequency of region-wide disasters as the result of an increasing number of especially violent storms. At some level, even a well-prepared Federal Emergency Management Agency (FEMA) system might be overwhelmed. As the cumulative magnitude of such damage increases, the federal government would likely leave state governments to shoulder more and more of the burden. The effect would be to strain the ligaments that hold the federal system together.

State governments are already pulling away from federal leadership on the environment. California is the leading example but others are coming along, mainly in the form of regional groupings.[4] The federal government is already fiscally compromised by defense costs in competition with escalating costs for maintaining the social contract. The additional costs entailed by climate change will make these problems unmanageable without drastic tradeoffs. At some point the government's ability to plan and act proactively will break down because the scale of events begins to overwhelm policies before they can generate appreciable results.

Western Hemisphere

Accumulated stresses owing to severe climate change may cause systemic economic and political collapse in Central and Latin America. The collapse of river systems in the western United States, for example, will also have a devastating effect on northern Mexico.[5] In Mexico, climate change likely will mean mass migration from central lowlands to higher ground. Immigration from Guatemala and Honduras into southern Mexico (whether for employment in Mexico or passage to the United States) is already a major issue for the Mexican government and will intensify dramatically. The pass-through consequence for the United States is that border problems will expand beyond the possibility of control, except by drastic methods and perhaps not even then. Efforts to choke off illegal immigration will have increasingly divisive repercussions on the domestic social and political structure of the United States.

Severe climate change will likely be the deathblow for democratic government throughout Latin America, as impoverishment spreads. In these circumstances we should expect that populist, Chavez-like governments will proliferate. Some regions will fall entirely and overtly under the control of drug cartels. Some governments will exist only nominally, and large regions will be essentially lawless, much as has been the case in Colombia. The United States will lack adequate means for responding effectively and will likely fall back on a combination of policies that add up to quarantine.

Tensions will increase between the United States and Canada, including clashes over fishing rights on both coasts. Two-thirds of Canadians rely on the Great Lakes, a relatively small watershed, for water, yet Great Lakes water levels are projected to decline by up to one foot in this century, attributable to increased evaporation and population growth. If the United States decides to divert water from the Great Lakes to compensate for the effects of climate change, the ingredients for a fundamental clash of interests with Canada would be in place. The opening of a Northwest Passage to shipping will also lead to an entirely new set of problems relating to navigation and resource rights. It cannot be excluded that Canada's tensions with the United States will play into domestic issues affecting the stability of Canada itself—most notably the western provinces' new role as oil exporters.

The cumulative effect of all these and related factors will be to render the United States profoundly isolated in the Western Hemisphere: blamed as a prime mover of global disaster, hated for self-protective measures it takes.

Europe and Eurasia

The prospect of a new ice age in Europe caused by the Gulf Stream's collapse is *not* an element of the severe climate scenario that serves as the basis for this chapter. But there is enough bad news for Europe in the scenario as it stands. Severe climate change will threaten every major port city in Europe and the United Kingdom. This will translate into huge economic costs at the national level and will prompt demands for EU intervention that are likely to exceed both its economic *and* its political resources. The Netherlands will be a particularly wrenching problem: a society at the core of European culture, which physically exists by restraining the sea, will be threatened by inundation. How will Europe share the costs of redesigning an entire nation?

Environmental pressures will accentuate the migration of peoples in numbers that will effectively change the ethnic signatures of major states and regions. In Europe, the influx of illegal immigrants from northern Africa and other parts of the continent will accelerate and become impossible to stop, except by means approximating a blockade. There will be political tipping points marked by the collapse of liberal concepts of openness, in the face of public demands for action to stem the tide. As the pressure increases, efforts to integrate Muslim communities into the European mainstream will collapse and extreme division will become the norm.

The beginnings of these trends are present already, but severe climate change will cause them to become far worse. One of the casualties of this process may be any prospect for the cultural, much less the political, integration of Turkey into the EU. Even if Turkey were to be admitted, the increasing reaction of Europeans against Islam may alienate the Turkish people, thereby destroying the hoped-for role of Turkey as a bulwark against radical Islam. At severe levels of climate change, civil disorder may lead to the suspension of normal legal procedures and rights. The precedents for dealing with large, unwanted minorities have already been set in Eurasia under fascism and communism. Under conditions marked by high levels of civil confusion and fear, political leaders and movements will emerge who might not resist these solutions.

In parts of the Russian Federation the Slavic population will continue receding while immigration from Asia intensifies. At some point these tensions may accumulate to the point where Moscow and Beijing collide over matters each believes to be vital to its own political stability and to the survival of its regime. Growing Asian settlement in portions of the Russian Federation will also result in increased friction, specifically with Russia's rapidly growing Islamic population.

The Russian core of the Federation will certainly not respond to these developments by shifting to liberal democracy. On the contrary, the antidemocratic legacy of the Putin period will be reinforced. Russia will return to its roots—to a czarist-type system in all but name, with the wealth of the country divided among members of a new "boyar" class as payment for loyalty. This regime will anchor itself ideologically in Russian nationalism and economically in the country's dominant energy position, which it will exploit aggressively. These trends are established already. Severe climate change will intensify them under Putin's successors.

Rising sea levels and accentuated storm systems will threaten China's industrialized coastal regions. Chinese economic growth will suffer as a result of the accelerated loss of land fertility resulting from the salinization of river deltas, which will compound the loss of arable land that has already occurred because of urbanization. Decreased rainfall will accelerate China's already critical shortage of water, not only for drinking but also for industrial purposes. This will also cancel out the promised effects of massive hydro-engineering projects such as the Three Gorges Dam.

There will be significant environmental pressures arguing for an inland shift of economic activity. China might be better able than other societies to accomplish this kind of transition, but the western reaches of China are water- and resource-poor. China will also find itself in direct confrontation with Japan and even the United States over access to fish, at a time when all major fisheries will likely have crashed as the result of today's unsustainable fishing practices, combined with the ongoing worldwide decimation of wetlands.

All this can place tremendous additional pressure on the national concept and on the Chinese political system. That system is already under stress; witness tens of thousands of clashes each year between the populace and local authorities. Political reform and liberalization of government control may be the necessary response to this kind of discontent, but severe climate change is much more likely to push China's central government, as well as the provincial governments, in the opposite direction.

Indian Subcontinent

On the Indian subcontinent the impact of global warming will be very destabilizing. As glaciers melt, the regions bounding the Indus and Ganges rivers will experience severe flooding. Once the glaciers are gone, the floods will be replaced by profound and protracted drought. The inland backflow of saltwater, caused by higher sea levels, will contaminate low-lying, fertile delta regions. Bangladesh, already famously vulnerable to storm surges, will become more so as sea levels rise.

Given the subcontinent's size and the variety of its regions, it is not possible to confidently interpolate from the IPCC's very broad findings down to the specifics needed for detailed political and security analysis. It is reasonable to say, however, that new and intense environmental pressures will be bad for the internal stability of all countries on the subcontinent, and bad for their relations with each other. At severe levels of climate change, the survival of Indian democracy will be at risk.

The Indus River system is the largest contiguous irrigation system on Earth with a total area of 20 million hectares (about 78,000 square miles) and an annual irrigation capacity of more than about 12 million hectares (about 45,000 square miles) in India, Bangladesh, Pakistan, and Nepal. The headwaters of the basin are in India; thus India is the most powerful player.[6] Currently, the other three countries of the system are engaged in water disputes with India. The Indus Water Treaty of 1960 settled some overarching issues, but frequent disagreements persist. (Pakistan now considers India in breach of the treaty for having caused "man-made river obstructions.")[7] Climate change will exacerbate these tensions. Because of India's clear upper hand, Pakistan may resort to desperate measures as it seeks water security.

North Africa and the Middle East

The northern tier of African countries will face collapse as water problems become unmanageable, particularly in combination with continued population growth. Morocco may be destabilized as a result of drought-induced failure of that country's hydroelectric power system and its irrigation-based agriculture. Those countries that can afford it may follow Libya's lead and attempt to tap major aquifers in a zero-sum struggle for survival. Muammar al-Qaddafi's $20 billion mass irrigation project would drain much of Great Nubian Sandstone Aquifer, which is nearly the size of Germany, within fifty years. Newly oil-rich Sudan is seeking to irrigate some of the Sahel; Ethiopia has claimed that any Sudanese effort to divert water from the Nile would provoke military response. Egypt will clash with Sudan or Ethiopia or both over any effort by either country to manipulate the flow of waters tributary to the Nile.

Efforts to design a solution to the Israeli-Palestinian struggle will be abandoned for the indefinite future because of a collective conclusion that the problem of sharing water supplies must be regarded as permanently intractable. War between Israel and Jordan over access to water is conceivable. Moreover, Iraq, Syria, and Turkey are likely to be enmeshed in an escalating struggle over the latter's command of waters feeding the Tigris and Euphrates systems. In the Gulf countries there will be a rapid expansion of nuclear power for desalinization. This in turn will become a contributing factor in the regional proliferation of nuclear weapons as insurance against predation.

Rising sea levels will cause extensive damage to delta regions—normally among the most fertile and heavily settled—as sea water presses farther upstream. This is already a problem in the Nile Delta, where the accelerated loss of fertile land will compound the impact of Egypt's oncoming demographic "youth bulge."

Sub-Sahara and the Horn of Africa

In sub-Saharan Africa, hundreds of millions of already vulnerable persons will be exposed to intensified threat of death by disease, malnutrition, and strife. Natural causes such as long-term drought will play a major role, but political factors either will exacerbate these disasters or may even precipitate them as the result of a mix of mismanagement and miscalculated policy. Such was the case in Ethiopia during the rule of Colonel Mengistu Haile Mariam. The ongoing genocide in Darfur may have begun as a consequence of water scarcity, as noted elsewhere in this report.

Under conditions of severe global climate change environmental factors will push already failed states deeper into the abyss, while driving other states toward the brink. The stronger regional states, such as South Africa, will be affected not only by internal social and economic stress related to changing climatic patterns but also by southward flows of refugees hoping for rescue and safety.

Contemporary Africa aspires to be a unified system but falls far short. Severe climate change would, in a grim way, provide for the first time the missing element of connectivity. From one end of the African continent to the other, severe climate change will become the common denominator of turbulence and destruction.

Systemic Events

As noted earlier, this chapter's analytic premise is that massive nonlinear events in the global environment will give rise to massive nonlinear societal events. The specific profile of these events will vary, but very high intensity will be the norm.

—We could see class warfare as the wealthiest members of every society pull away from the rest of the population, undermining the morale and viability of democratic governance, worldwide.

—It is possible that global fish stocks will crash. Signs are that this process is already well established and accelerating. Aquaculture will expand dramatically to mitigate fish protein shortages, but the destruction of natural marine food chains will have an incalculable impact on the viability of the oceans themselves.

—Climate change may have serious impacts on disease vectors. Under conditions of extreme climate change the risk of pandemic explosions of disease increase.

—As drinkable water becomes scarcer it will become an increasingly commercialized resource. Governments, lacking the necessary resources, will privatize supply. Experience with privatized water supply in poor societies suggests the likelihood of violent protest and political upheaval.

—Human fertility may collapse in economically advanced regions, as the consequence of increasingly difficult living conditions and of general loss of hope for the longer term.

—Globalization may end and rapid economic decline may begin, owing to the collapse of financial and production systems that depend on integrated worldwide systems.

—Corporations may become increasingly powerful relative to governments as the rich look to private services. This may engender a new form of globalization in which transnational business becomes more powerful than states.

—Alliance systems and multilateral institutions may collapse—among them the UN, as the Security Council fractures beyond compromise or repair.

Moral Consequences

Massive social upheaval will be accompanied by intense religious and ideological turmoil, as people search for relief and hope. For this purpose, it is fair to consider that certain kinds of political doctrine may be thought of as religious. Fascism and communism certainly filled that role for true believers during the twentieth century. Among traditional religious beliefs, the "losers" are likely to be those faiths that have formed the closest associations with the secular world and with scientific rationalism. Among political systems, authoritarian ideologies would certainly be the "winners." One way or the other, severe climate change will weaken the capacity of liberal democratic systems to maintain public confidence.

This intensified search for spiritual meaning will be all the more poignant under conditions of severe climate change. Governments with resources will be forced to engage in long, nightmarish episodes of triage: deciding what and who can be salvaged from engulfment by a disordered environment. The choices will need to be made primarily among the poorest, not just abroad but at home. We have already previewed the images, in the course of the organizational and spiritual unraveling that was Hurricane Katrina. At progressively more extreme levels, the decisions will be increasingly harsh: morally agonizing to those who must make and execute them—but in the end, morally deadening. For comparison one might look to estimates of the effects of a new global pandemic carried by avian flu.

Die-off

War and disease can be the means to achieve a grim kind of environmentally sustainable relationship between humankind and nature. Hundreds of millions of people already survive on a hand-to-mouth basis, living essentially on the leavings and limited charity of those who are better off. As climate change deepens, even the donor portion of society will feel the effects, and those below will be much worse off than before.

Severe climate change will put additional stress on all systems of social support. Already tenuous health care systems may collapse. Vulnerability to new forms of disease will increase. In some regions the process may resemble the abrupt die-offs that are thought to have occurred on a smaller scale among ancient peoples. Instead of focusing on ways to save modern civilization, social efforts may increasingly focus on sheer survival. Preemptive desertion of urban civilization will occur. Attention to the long-term requirements of society will weaken as a consequence of a public conviction that nothing can be done to alter the downward course of events.

Survival and Reconstruction

The consequences of even relatively low-end global climate change include the loosening and disruption of societal networks. At higher ranges of the spectrum, chaos awaits. The question is whether a threat of this magnitude will dishearten humankind or cause it to rally in a tremendous, generational struggle for survival and reconstruction.

If that rally does not occur relatively early on, then chances increase that the world will be committed irrevocably to severe and permanent global climate change at profoundly disruptive levels. An effective response to the challenge of global warming cannot be spread out across the next century, but rather must be set in place in the next decade, in order to have any chance to meaningfully alter the slope of the curves one sees in the IPCC report. We are already in the midst of choosing among alternative futures. The onset of these choices is rapid, and the consequences are likely to be irreversible. Moreover, the upper end of the "severe, thirty-year scenario" can just as well be a prelude to even worse circumstances, if the political will to deal with global warming collapses early on under the weight of universal pessimism.

In order to emerge from a period of severe climate change as a civilization with hopes for a better future and with prospects for further human development, the very model of what constitutes happiness must change. Globalization will have to be redirected. It cannot continue forever in its present form, based on an insatiable consumption of resources. The combined demands of China and India alone cannot be satisfied in a world already heavily burdened by the consumption patterns of the United States, Europe, and Japan.

Levels of demand will have to be brought into line with the availability of resources. This can occur either as the result of the collapse of the present system or by its purposeful reconfiguration. Yet neither China nor India can voluntarily accept that their hopes for full-fledged consumer societies cannot be realized. The promise that it is possible to achieve high levels of consumption for all people everywhere would remain unfulfilled. The ideal of international development would be seen to have failed, with profound political consequences.

Deus ex machina?

It is possible that science and engineering—supported at levels commensurate with recognition of a climate change as a global, existential threat—will produce solutions powerful enough to redirect the history of the future.

Internet

The Internet continues to penetrate and transform all spheres of activity. It represents one of the few binding forces that may not only survive severe climate change, but continue to develop. For this to happen, the Internet must remain open as a channel for open scientific exchange and policy debate, especially as other avenues of communication are badly affected by growing social and international tensions. Efforts by some governments to muzzle the Internet, and efforts by large corporations to dominate it, will impinge on its value as a means for allowing civilization to meet and deal with global warming.

Extreme Technologies

Today, we talk about expanding the use of conservation, and of nuclear power, or about intensified use of renewable sources of energy such as solar or wind or wave power. These technologies dampen the trajectory of climate change at lower levels. Under conditions of severe climate change, however, more radical solutions would be needed. A wish-list would include practical fusion power; the full blossoming of nanotechnology as a source of new materials with radically superior properties; a full, closed-cycle approach to the use of resources; global-scale carbon sequestration; bioengineered plant and animal life, capable of providing for the food supply under harsh climate conditions; and full-scale planetary engineering.

Climate Change Science and Climate Change Policy

In the field of climate change, one can see an ongoing transformation in the relationship between science and policy. Normally, the scientific community is reticent about theory, hedging its conclusions in the face of unknowns and unpredictable variables. The IPCC studies mark the point where much of the scientific world has lost that hesitance, and instead has closed ranks behind three conclusions: first, that climate change is in process; second, that regardless of underlying cyclical trends, the largest part of the change is induced by human activities; and third, that it is up to political and economic leaders to take action commensurate to the potential, rather than the demonstrated, scale of the change and its consequences.

Policymakers can no longer take cover behind scientific disagreements as to the reality of global warming or its causes, although some will still try to cherry-pick dissident opinion. Until now, a case could be made that policy should move very slowly in response to scientific warnings about climate change, given the scale of what would have to be done. Delay was prudent, and prudence is good. Now, in the face of the massive IPCC consensus, protracted delay is tantamount to reckless neglect of the public trust.

Uncertainties Regarding Specifics of Impact

The dominant scientific view is that global warming is real, and that its cause is human activity. The conversion of this knowledge into policy analysis, on the other hand, is not a clear-cut process, as illustrated by the following three passages:

> What we currently have some idea about is the relative vulnerability of various locations to changes in sea level, based on coastal elevation, population density, coastal agricultural production, and ability to defend against rising seas (e.g., the Netherlands is relatively less vulnerable despite its low elevation than, say, the Mekong Delta, which has a much larger population density, lower income, few existing defenses, and a large amount of Viet Nam's cereal production). Given this type of background, we can simply say who is more likely to be damaged, and assume that the higher the sea level the more damage there is to deal with (or not).[8]

> There has not been a systematic study of the problem of when small island cultures collapse on the whole. It also appears that this will vary geographically. . . . The timing of sea level rise will vary geographically,

> regardless of how fast the average rate is. It is very possible that sea level contributed from polar ice sheets will occur in decade-long pulses with lulls in between. This could put the most vulnerable areas in danger suddenly and without warning. It could also overrun attempts to build protective structures for major coastal cities, where adaptation is underway but major public works projects are "surprised" by unanticipated rapid rise. Note that we do not know how sea level rise will vary geographically. The pattern of differences observed in the past is not thought to indicate how the pattern will be in the future. . . . Models disagree widely on the future pattern.[9]

> The diversity of African climates, high rainfall variability, and a very sparse observational network make predictions of future climate change difficult at the subregional and local levels. Underlying exposure and vulnerability to climatic changes are well established. Sensitivity to climatic variations is established but incomplete. However, uncertainty over future conditions means that there is low confidence in projected costs of climate change.[10]

Front-Loaded Policy

The security implications of extreme climate change depend in part not only on the estimated consequences but also on the political impact of those estimates and of ensuing facts on the ground. How well our civilization rallies to meet the threat of extreme climate change will depend in large part on public confidence in government's ability to minimize the damage. That confidence is likely to be greater if governments move sooner and faster to deal with these events than if they are seen to have lost the best chance through delay. Conversely, if governments fail in this responsibility, public expectations of future success can collapse. The severe climate change scenario that is the subject of this chapter inherently represents multiple failures: failure of science to make correct assessments early on, and failure of policy to have acted upon earlier warnings. It is a sequence of events that would deeply undermine public confidence in the possibility of effective action.

Radical Change in Time-Horizon

Climate change represents a permanent, not a reversible, shift in the relationship of humankind to nature. Since we already have attained the power to permanently interfere with natural cycles, we are now permanently in charge of regulating our impact upon them. Designing an approach for this is daunting, but it is not beyond our capabilities. We are already confronting—in bits and pieces—the reality of having to permanently manage environmental systems. Radioactive waste requires disposal methods that will last for half-lives measured in scores of thousands of years. Genetically altered plants and animals can produce permanent changes in the biosphere.

Environmental Intelligence: MEDEA

Policymakers will need to know that all available information is being used to help guide assessments of global warming and the political dynamics it will generate. This suggests an important role for intelligence resources, both databases and analytic talent.

In the 1990s there was a serious effort, funded by Congress, to bring about a sustained engagement between top environmental scientists and the U.S. intelligence community. This process, known as MEDEA, was gradually defunded for political reasons, and then eliminated. As a result of MEDEA's activities, thirty years of Corona satellite photography were declassified and released from the archives, for use by environmental scientists in establishing ground truth for terrestrial change over time; U.S. and Russian data on Arctic ice thickness were compiled and jointly released to scientists looking for evidence of ice thinning resulting from global warming; a freestanding environmental intelligence unit was created on an interagency basis; and detailed plans for a tsunami warning system (as part of the GDIN—Global Disaster Information Network) were developed—though not engaged until after the disastrous event of December 2004.

We clearly need MEDEA or something much like it. Environmental issues, especially those driven by climate change, are already national security issues. To deal with such issues, new forms of current and longer-range intelligence will be required. New kinds of international environmental agreements will require new monitoring systems. Some of these agreements may involve reprocessing data from existing collection platforms, including data platforms that are under civil control. Others may require investment in new sensors, either as add-ons to projected collection systems or as entirely new packages. A new analytic system will surely be needed to keep track of the societal consequences of climate change as it advances.

The National Intelligence Council is said to be preparing a National Intelligence Estimate on climate change. One-off assessments of this kind are vital, but we especially need an ongoing analytic process. A new field of intelligence analysis must be brought into being—and it would not be the first time this has happened. A cross section of the intelligence world would show a series of "growth rings" corresponding to periods of time when investments were made to develop intelligence that informs capabilities to respond to newly recognized challenges: counterinsurgency; drug-trafficking measures; counterproliferation and counterterrorism measures; and now intelligence for the age of global climate change.

Restoring and revising MEDEA in an updated form would help to recouple the scientific community to this intelligence effort. In the end, however, the capacity to integrate environmental data from all sources is one that should be undertaken not only by the intelligence community but by the government as a whole, linking and combining all assets for assembling and processing data bearing on climate change and on the interaction between climate change and policy.

Managing Complexity

Complex phenomena such as extreme global climate change have certain hallmarks: they represent the effects of a system in motion, wherein all parts influence and are influenced by all other parts, more or less concurrently, in a permanently ongoing process. Responses to such problems require an understanding of their dynamics, including an awareness that actions targeted to any part of the system will generate consequences throughout the whole system.

This new class of problems tends to be fast moving and unstable, in the sense that trends and events interact spontaneously, with the result that unforeseen developments can outpace societal response. In complex systems, inputs and outputs are not only unpredictable but will on occasion be highly nonlinear: that is, seemingly small events will lead to massively consequential results. Complex systems do not tend toward stability and in fact harbor the possibility of collapse. There are no permanent solutions for problems arising out of complexity: instead, problems mutate and require permanent management.

These qualities present a severe challenge to our ingrained approach to governance. We artificially distinguish between domestic and international policy, although there is no way to understand either in isolation. We discount the impact of future events in favor of immediate concerns, ignoring the ramifications of our short-term decisions for our longer-term interests. Despite our awareness of the need for integration of all policy areas, the formation and execution of policy are poorly articulated. Despite the fact that issues mutate continuously, the executive branch and Congress habitually tout policies as perfect, permanent solutions.

"Robust" Options

Under the circumstances, policymakers might justifiably conclude that the search for long-range approaches to climate change is hopeless, beyond human foresight and capacity. Fortunately, there are "workarounds," in the form of computer-assisted modeling designed to sift vast numbers of possible fits between circumstances and adaptive policies. The Rand Corporation's Frederick S. Pardee Center for Longer Range Global Policy has been pioneering this approach, which involves what center scientists have termed the search for "robust options."

Robust options are not perfectly fine-tuned to one and only one set of ideal conditions. Such options are in fact weeded out because they are very unforgiving and have a tendency to collapse disastrously. Robust options are intended to tolerate a range of less-than-optimal conditions. Computers enter the picture by making it possible to change parameters at will, thousands of times, if needed. It is hoped that this technique will enable policymakers to gain an edge in their battle against complexity and chaos.

Networking Governance

Today's issues—such as climate change—are already complex, yet today's governance is largely based on an outmoded hierarchical model. We require systems of organization for government as a whole that promote earlier detection and response to both opportunity and error; earlier alertness to interactions across substantive boundaries; and the ability to organize and apply long-range foresight as a new and crucial dimension of governance.

Networking is society's best organizational response to complexity. It captures the idea of a flattened form of organizational network, where coherence occurs because of the presence of the following:

—A strategic concept, clearly articulated by senior leadership and understood at all levels

—Information flows that support initiative at lower levels

—Feedback at every level to allow continuing interaction and situational awareness within the group

—Something very like complex adaptive behavior toward realization of the goal (mission)

—A culture within the organization such that its members are encouraged to self-organize, rather than await instruction from above[11]

Networking is a form of social action that is profoundly well suited for democratic governance. Networking depends upon the existence of widely distributed intelligence and initiative. It also depends on the existence of a collaborative ethos, whereby an instinct for teamwork operates as an offset to the natural search for individual advantage. Democracy has always required faith that these qualities would serve social as well as individual needs. It has also always required farsighted investment in the education of individuals for their role as citizens.[12]

This chapter is not the place for a discussion of how to design the necessary changes in our executive, legislative, regulatory, and political systems.[13]

Global Governance

The IPCC process shows that global scientific cooperation is possible and that it can be highly effective in dealing with complex scientific problems. However, cooperation on issues of complex policy is another matter. The Ozone Convention shows that effective worldwide action can be based on timely scientific warning, but global warming is a system of issues that is orders of magnitude more difficult to manage than eliminating CFCs (chlorofluorocarbons). Meeting this challenge will require an unprecedented level of sustained international collaboration.

There will be talk of the need for global government. It is unlikely, however, that such a government will emerge. That is just as well, since such a system would tend to accumulate power at the expense of freedom, without a proportionate increase of wisdom. A form of flexible global governance, on the other hand, is perhaps the best available outcome. It would not be linear and orderly, but "fuzzy" and networked. NATO was such a system: it worked against the threat of war in Europe, and it continues to be relevant. The prospect of severe climate change may be what it takes to bring into existence another grand alliance, designed to deal with climate change. In any effort to define such an approach, U.S. political leadership will be essential.

The Challenge to American Leadership

The United States' lasting contribution to the world is not just the idea of popular democracy but the ongoing demonstration that it works: that free citizens have not only the inherent right to govern themselves but also the capacity to do so. These claims have been tested in every period of our national existence. Today, they are challenged by an accelerating rate of profound social change, wherein powerful domestic and global forces continuously and vigorously interact. Climate change is one of the most potent of these forces—already viewed by some as the central crisis of our era. When civil governance fails to deal with crisis, there are default options that might otherwise seem unthinkable. Climate change will produce a succession of concurrent crises. The aftermath of Hurricane Katrina, which saw a public clamor for military intervention to replace a disorganized and inept civilian presence, is a cautionary tale. Legitimacy is based, in the end, on performance.

Postscript: Need for a Sustained Analytic Effort

Guesses about the political and security consequences of extreme climate change are problematic, but the questions they raise are not. They represent an urgent agenda for analysts and policymakers, to be undertaken now and continued indefinitely. As we react to global warming, we will inevitably change it and its consequences. Policies good enough to launch the effort will eventually lose their value. Unintended consequences will suddenly appear, and eventually be recognized—the sooner, the better. The one constant will be change. And the one indispensable response to climate change must be the will to learn and to adapt.

Conclusion

The reduction of humankind's burden on the environment can occur as the result of deteriorating physical conditions and attendant pandemics. It can also occur as the result of war and its aftermath. Under the circumstances described earlier, it is clear that even nuclear war cannot be excluded as a political consequence of global warming—and so-called "limited nuclear war" in any part of the world can escalate to a full-scale nuclear exchange among the big nuclear powers. Even if one assumes that there will be very large reductions of nuclear weapons in the inventories of the United States and the Russian Federation, it should be kept in mind that the weapons on board a *single* submarine armed with ballistic missiles are fully capable of destroying a nation of continental size.

The alternative to reducing populations by decimation is to reduce them by demographic management. Every nation has a demographic curve, showing the rate at which the size and composition of its population will change over time, given certain assumptions. Today, advanced states use macroeconomic techniques to manage their economies; tomorrow, such states may be looking for macro-techniques to manage reproductive choice to meet basic targets. This is a radical departure, given the way people everywhere feel about reproductive freedom. But if the alternative is truly ruinous, what is currently unthinkable may wind up on the table. China will be an early bellwether.

Climate change represents a permanent shift in the relationship of humankind to nature. Since we already have attained the power to alter natural cycles, we are now accountable for regulating our impact upon them. To fulfill this stewardship responsibly we must improve the capacity of governance to deal with all kinds of complex phenomena: through earlier recognition and response to important challenges; deeper awareness of interactions across substantive and bureaucratic boundaries; and the ability to organize and execute policy for operation over extended periods of time. Finding and applying the necessary political and governmental innovations is daunting, but it is a task within our capabilities, as has been repeatedly demonstrated in the course of human history.

Notes

1. IPCC (Intergovernmental Panel on Climate Change), *Climate Change 2007: Impacts, Adaptation and Vulnerability: Contribution of Working Group II to the Fourth Assessment Report of the Intergovernmental Panel on Climate Change,* edited by M. L. Parry and others (Cambridge University Press, 2007) (www.ipcc.ch).

2. C. Rosenzweig, "Using Regional Models to Assess the Potential for Extreme Climate Change" (New York: Columbia University Center for Climate Systems Research, 2004).

3. "Ogallala Aquifer," *Encyclopedia of Water Resources* (www.waterencyclopedia.com/Oc-Po/Ogallala-Aquifer.html).

4. Emma Marris, "Western States Reach Carbon Scheme," *Nature* 446 (2007): 114–15.

5. John Opie, *Ogallala: Water for a Dry Land* (University of Nebraska Press, 1993).

6. Chietigj Bajpaee, "Asia's Coming Water Wars," *Power and Interest News Report,* August 22, 2006.

7. Ibid.

8. Jay Gulledge, Pew Center on Global Climate Change, personal communication, April 28, 2007.

9. Ibid.

10. IPCC, *Climate Change 2001: Synthesis Report: Third Assessment Report of the Intergovernmental Panel on Climate Change,* edited by Robert T. Watson (Cambridge University Press, 2001).

11. David S. Alberts and Richard E. Hayes, *Power to the Edge: Command and Control in the Information Age* (Vienna, Va.: Command and Control Research Program Publications, June 2003) (free download: www.dodccrp.org/publications/pdf/Alberts_Power.pdf).

12. Leon Fuerth, "Congress and the Climate Crisis: A Case for Forward Engagement," Research Brief 3 (New York University, Brademas Center for the Study of Congress, 2007).

13. There is a discussion of topics such as this in my work as a research professor at George Washington University, and as director of the Forward Engagement project (financed by the Rockefeller Brothers Fund and George Washington University). For further discussion, see www.forwardengagement.org.

six

SECURITY IMPLICATIONS OF CLIMATE SCENARIO 3

Catastrophic Climate Change over the Next Hundred Years

SHARON E. BURKE

> Twenty years we've had the drought / And our reservoirs have all dried up
> I take my baths now in a coffee cup / I boil what's left of it for tea.
> ——*Urinetown, The Musical*

Two years after Hurricane Katrina, New Orleans is still not a fully functioning city, despite the expenditure of billions of dollars and the attention of a nation. As of November 2007, population levels were only 70 percent of the pre-hurricane levels and nearly 47,000 families continued to live in FEMA trailers. Sixty-two percent of the schools and 38 percent of the day care facilities had reopened, and only 19 percent of public buses were running. Only 36 percent of the people who had applied for "Road Home" grants to help them rehabilitate their properties had received funding.[1] With social, public, and criminal justice systems in disarray, long-standing criminal and corruption problems have exploded, making New Orleans the murder capital of America.[2]

Now imagine another scenario: Hurricane Katrina hit New Orleans as a category 5 storm, instead of a 3. Or consider the possibility that another hurricane could have made landfall in the years since. How many more people would have perished in the storms? How much more of the city would have been destroyed? Would communities such as Baton Rouge and Houston have been able and willing to absorb larger numbers of storm victims, on a permanent basis? It is hard to speculate exactly what would have happened, but it is very clear that any progress made in reconstructing New Orleans after Hurricane Katrina would have been seriously damaged by any subsequent storm. Perhaps the city would have remained uninhabitable.

In the catastrophic climate scenario that this study analyzes, that could be the future for Jacksonville and Houston, and also for New York City and Washington, D.C. It could be the future for cities all over the world, if not from a hurricane, then because of massive sea level rise or prolonged drought, or relentless summer heat waves, or any number of other consequences of unprecedented climate change.

In this long-range scenario, intense hurricanes may become increasingly common, and droughts, floods, wildfires, heat waves, and churning seas certainly will. Hundreds of millions of thirsty and starving people will have to flee or perish, leaving the globe dotted with ghost towns. The abrupt and sudden nature of many of these phenomena will challenge the ability of all societies to adapt, including the United States. Persistent conflict—civil, communal, sectarian, regional, and between nations—will be the norm in this plausible scenario. Indeed, even this scenario may be too conservative.

It is very difficult to imagine such a grim reality. A world 5.6°C (10.1°F) warmer has not existed for more than 40 million years, and sea level 2 meters (6.6 feet) higher has not existed for more than 100,000 years. Today, the world has entered the Information Age, which may reshape the foundation of societies around the world as profoundly as did the Industrial Age. There certainly are grave threats in this increasingly decentralized, engineered, and networked world, but at the same time this is overwhelmingly a time of hope and bright horizons as millions of people are lifted out of poverty and empowered. Moreover, this is a transition in its infancy—where it will take human society in the years ahead and how fast is unclear.

Over the course of the next century, however, if the catastrophic scenario described in this chapter comes to pass, these hopes will be eclipsed as all nations on the Earth struggle to meet the challenges of profound climate change. In this scenario, by the end of this century, the world will have entered the Age of Survival.

The Risks of Prediction

In 1902 the writer H. G. Wells predicted that "long before the year A.D. 2000, and very probably before 1950, a successful aeroplane will have soared and come home safe and sound" and he predicted such a device would become an important weapon in warfare. He also suggested that in the year 2000, cooks would no longer labor with "crimsoned faces and blackened arms" over open fires, but rather over "a neat little range, heated by electricity and provided with thermometers, with absolutely controllable temperatures and proper heat screens."[3]

Many of Wells's predictions for fifty to one hundred years in the future—the rise of cities; the ubiquity of electricity; mechanized mass transportation; highly destructive world wars—were remarkably prescient, given that Thomas Edison had invented the light bulb only twenty years before, the Wright Brothers' first successful flight had yet to occur, and the Model T was still seven years away. But Wells was hardly perfect. He also predicted the end of democracy and the rise of dirigibles. A *Ladies Home Journal* article of predictions for the year 2000, published around the same time, accurately predicted the population of the United States, but also foresaw that "strawberries as large as apples will be eaten by our great-great-grandchildren."[4]

Forecasting the distant future has always been a risky but necessary part of organizing how we think and act in the present. In general, our visions for the distant future are limited by what we already know and are strongly grounded in the past. Electricity and dirigibles were recent innovations when H. G. Wells wrote his predictions, so describing the electrification of daily life and strategic importance of blimps in the future made sense.

Today, however, it is very difficult to base views of the distant future on the record of the past and an understanding of the present for two reasons. The first, of course, is climate change. This is an inexorable trend of unprecedented global scope, with an enormous range of uncertainty. The plausible effects, from repeated, severe weather events to prolonged, record high temperatures, are unprecedented in human history. There may be wholly new climate zones, and some of the temperature and precipitation patterns will be entirely new.[5] We can base our predictions on exceptional events, such as Hurricane Katrina or the 2003 heat wave in Europe, but there is no true record to inform our views of the future.

The speed of technological change today is another unprecedented and global trend. If the present rate of change continues, computers will approximate the power of the human brain within twenty-five years; twenty-five years after that, one computer "could have the processing power of all human brains."[6] What that might mean is the province of science fiction writers. Nanotechnology, the construction and manipulation of materials at a molecular level, may radically change human society, and bioengineering may even redefine what it means to be human. It is extremely difficult to comprehend what these radical trends will mean for the future, particularly given that the pace of change itself has increased exponentially.[7]

At the same time, however, the implications of both trends for human society and survival raise the stakes; it is crucial to try to understand what the future might look like in one hundred years in order to act accordingly today. This scenario, therefore, builds a picture of the plausible effects of catastrophic climate change, and the implications for national security, on the basis of what we know about the past and the present. The purpose is not to "one up" the previous scenarios in awfulness, but rather to attempt to imagine an unimaginable future that is, after all, entirely plausible.

Assumed Climate Effects of the Catastrophic Scenario

In the catastrophic scenario, the year 2040 marks an important tipping point. Large-scale, singular events of abrupt climate change will start occurring, greatly exacerbated by the collapse of the Atlantic meridional overturning circulation (MOC), which is believed to play an important role in regulating global climate, particularly in Europe.[8] There will be a rapid loss of polar ice, a sudden rise in sea levels, totaling 2 meters (6.6 feet), and a temperature increase of almost 5.6°C (10.1°F) by 2095.

Developing countries, particularly those at low latitudes and those reliant on subsistence, rain-fed farming, will be hardest and earliest hit. All nations, however, will find it difficult to deal with the unpredictable, abrupt, and severe nature of climate change after 2040. These changes will be difficult to anticipate, and equally difficult to mitigate or recover from, particularly as they will recur, possibly on a frequent basis.

First, the rise in temperatures alone will present a fundamental challenge for human health. Indeed, even now, about 250 people die of heatstroke every year in the United States. In a prolonged heat wave in 1980, more than 10,000 people died of heat-related illnesses, and between 5,000 and 10,000 in 1988.[9] In 2003, record heat waves in Europe, with temperatures in Paris hitting 40.4°C (104.7°F) and 47.3°C (116.3°F) in parts of Portugal, are estimated to have cost more than 37,000 lives; in the same summer there were at least 2,000 heat-related deaths in India.

Average temperatures will increase in most regions, and the western United States, southern Europe, and southern Australia will be particularly vulnerable to prolonged heat spells. The rise in temperatures will complicate daily life around the world. In Washington, D.C., the average summer temperature is in the low 30s C (high 80s F), getting as high as 40°C (104°F). With a 5.6°C (10.1°F) increase, that could mean temperatures as high as 45.6°C (114.5°F). In New Delhi, summer temperatures can reach 45°C (113°F) already, opening the possibility of new highs approaching

50.5°C (123°F). In general, the level of safe exposure is considered to be about 38°C (100°F); at hotter temperatures, activity has to be limited and the very old and the very young are especially vulnerable to heat-related illness and mortality. Sudden shifts in temperature, which are expected in this scenario, are particularly lethal.

As a result of higher temperatures and lower, unpredictable precipitation, severe and persistent wildfires will become more common, freshwater will be more scarce, and agricultural productivity will fall, particularly in Southern Europe and the Mediterranean, and the western United States. The World Health Organization estimates that water scarcity already affects two-fifths of the world population—some 2.6 billion people. In this scenario, half the world population will experience persistent water scarcity. Regions that depend on annual snowfall and glaciers for water lose their supply; hardest hit will be Central Asia, the Andes, Europe, and western North America. Some regions may become uninhabitable due to lack of water: the Mediterranean, much of Central Asia, northern Mexico, and South America. The southwestern United States will lose its current sources of fresh water, but that may be mitigated by an increase in precipitation due to the MOC collapse, though precipitation patterns may be irregular. Regional water scarcity will also be mitigated by increases in precipitation in East Africa and East and Southeast Asia, though the risk of floods will increase. The lack of rainfall will also threaten tropical forests and their dependent species with extinction.

Declining agricultural productivity will be an acute challenge. The heat, together with shifting and unpredictable precipitation patterns and melting glaciers, will dry out many areas, including today's grain-exporting regions. The largest decreases in precipitation will be in North Africa, the Middle East, Central America, the Caribbean, and northeastern South America, including Amazonia. The World Food Program estimates that nearly 1 billion people suffer from chronic hunger today, almost 15 million of them refugees from conflict and natural disasters. According to the World Food Program, "More than nine out of ten of those who die [of chronic hunger] are simply trapped by poverty in remote rural areas or urban slums. They do not make the news. They just die." Mortality rates from hunger and lack of water will skyrocket over the next century, and given all that will be happening, that will probably not make the news, either—people will just die.

Over the next one hundred years, the "breadbasket" regions of the world will shift northward. Consequently, formerly subarctic regions will be able to support farming, but these regions' traditionally small human populations and lack of infrastructure, including roads and utilities, will make the dramatic expansion of agriculture a challenge. Moreover, extreme year-to-year climate variability may make sustainable agriculture unlikely, at least on the scale needed. Northwestern Europe, too, will see shorter growing seasons and declining crop yields because it will actually experience colder winters, due to the collapse of the MOC.

At the same time that the resource base to support humanity is shrinking, there will be less inhabitable land. Ten percent of the world population now lives in low-elevation coastal zones (all land contiguous with the coast that is 10 meters or less in elevation) that will experience sea level rises of 6.6 feet (2 meters) in this scenario and 9.8 feet (3 meters) in the North Atlantic, given the loss of the MOC. Most major cities at or near sea level have some kind of flood protection, so high tides alone will not lead to the inundation of these cities. Consider, however, that the combined effects of more frequent and severe weather events and higher sea levels could well lead to increased flooding from coastal storms and coastal erosion. In any case, there will be saltwater intrusion into coastal water supplies, rising water tables, and the loss of coastal and upstream wetlands, with impacts on fisheries.

The rise could well occur in several quick pulses, with relatively stable periods in between, which will complicate planning and adaptation and make any kind of orderly or managed evacuation unlikely. Inundation plus the combined effects of higher sea levels and more frequent tropical storms may leave many large coastal cities uninhabitable, including the largest American cities, New York City and Los Angeles, focal points for the national economy with a combined total of almost 33 million people in their metropolitan areas today. Resettling coastal populations will be a crippling challenge, even for the United States.

Sea level rises also will affect food security. Significant fertile deltas will become largely uncultivable because of inundation and more frequent and higher storm surges that reach farther inland. Fisheries and marine ecosystems, particularly in the North Atlantic, will collapse.

Locally devastating weather events will be the new norm for coastal and mid-latitude locations—wind and flood damage will be much more intense. There will be frequent losses of life, property, and infrastructure—and this will happen *every year.* Although water scarcity and food security will disproportionately affect poor countries—they already do—extreme weather events will be more or less evenly distributed around the world. Regions affected by tropical storms, including typhoons and hurricanes, will include all three coasts of the United States; all of Mexico and Central America; the

Caribbean islands; East, Southeast and South Asia; and many South Pacific and Indian Ocean islands. Recent isolated events when coastal storms made landfall in the South Atlantic, Europe, and the Arabian Sea in the last few years suggest that these regions will also experience a rise in the incidence of extreme storms.

In these circumstances, there will be an across-the-board decline in human development indicators. Life spans will shorten, incomes will drop, health will deteriorate—including as a result of proliferating diseases—infant mortality will rise, and there will be a decline in personal freedoms as states fall to anocracy (a situation where central authority in a state is weak or non-existent and power has devolved to more regional or local actors, such as tribes) and autocracy.

The Age of Survival: Imagining the Unimaginable Future

If New Orleans is one harbinger of the future, Somalia is another. With a weak and barely functional central government that does not enjoy the trust and confidence of the public, the nation has descended into clan warfare. Mortality rates for combatants and noncombatants are high. Neighboring Ethiopia has intervened, with troops on the ground in Mogadishu and elsewhere, a small African Union peacekeeping force is present in the country, and the United States has conducted military missions in Somalia within the last year, including air strikes aimed at terrorist groups that the United States government has said are finding safe haven in the chaos.[10] In a July 2007 report, the UN Monitoring Group on Somalia reported that the nation is "literally awash in arms" and factional groups are targeting not only all combatants in the country but also noncombatants, including aid groups.

Drought is a regular feature of life in Somalia that even in the best of times has been difficult to deal with. These are bad times, indeed, for Somalia, and the mutually reinforcing cycle of drought, famine, and conflict has left some 750,000 Somalis internally displaced and about 1.5 million people—17 percent of the population—in dire need of humanitarian relief. The relief is difficult to provide, however, given the lawlessness and violence consuming the country. For example, nearly all food assistance to Somalia is shipped by sea, but with the rise of piracy, the number of vessels willing to carry food to the country fell by 50 percent in 2007.[11] Life expectancy is forty-eight years, infant mortality has skyrocketed, and annual per capita GDP is estimated to be about six hundred dollars. The conflict has also had a negative effect on the stability of surrounding nations.

In the catastrophic climate change scenario, situations like that in Somalia will be commonplace: there will be a sharp rise in failing and failed states and therefore in intrastate war. According to International Alert, there are forty-six countries, home to 2.7 billion people, at a high risk of violent conflict as a result of climate change. The group lists an additional fifty-six nations, accounting for another 1.2 billion people, that will have difficulty dealing with climate change, given other challenges.[12] Over the next hundred years, in a catastrophic future, that means there are likely to be at least 102 failing and failed states, consumed by internal conflict, spewing desperate refugees, and harboring and spawning violent extremist movements. Moreover, nations all over the world will be destabilized as a result, either by the crisis on their borders or the significant numbers of refugees and in some cases armed or extremist groups migrating into their territories.

Over the course of the century, this will mean a collapse of globalization and transnational institutions and an increase in all types of conflict—most dramatically, intrastate and asymmetric. The global nature of the conflicts and the abruptness of the climate effects will challenge the ability of governments all over the world to respond to the disasters, mitigate the effects, or to contain the violence along their borders. There will be civil unrest in every nation as a result of popular anger toward governments, scapegoating of migrant and minority populations, and a rise in charismatic end-of-days cults, which will deepen a sense of hopelessness as these cults tend to see no end to misery other than extinction followed by divine salvation.

Given that the failing nations account for half of the global population, this will also be a cataclysmic humanitarian disaster, with hundreds of millions of people dying from climate effects and conflict, totally overwhelming the ability of international institutions and donor nations to respond. This failure of the international relief system will be total after 2040 as donor nations are forced to turn their resources inward. There will be a worldwide economic depression and a reverse in the gains in standards of living made in the twentieth and early twenty-first centuries.

At the same time, the probability of conflict between nations will rise. Although global interstate resource wars are generally unlikely,[13] simmering conflicts between nations, such as that between India and Pakistan, are likely to boil over, particularly if both nations are failing. Both India and Pakistan, of course, have nuclear weapons, and a nuclear exchange is possible, perhaps likely, either by failing central governments or by extremist and ethnic groups that seize control of nuclear weapons. There will also be competition for the Arctic region, where natural resources, including oil and arable land, will be increasingly accessible and borders are ill defined. It is possible that agreements over Arctic territories will be worked out among Russia, Canada, Norway, the United States, Iceland, and Denmark in the next two decades, before the truly catastrophic climate effects manifest themselves in those nations. If not, there is a strong probability of conflict over the Arctic, possibly even armed conflict. In general, though, nations will be preoccupied with maintaining internal stability and will have difficulty mustering the resources for war. Indeed, the greater danger is that states will fail to muster the resources for interstate cooperation.

Finally, all nations are likely to experience violent conflict as a result of migration patterns. There will be increasingly few arable parts of the world, and few nations able to respond to climate change effects, and hundreds of millions of desperate people looking for a safe haven—a volatile mix. This will cause considerable unrest in the United States, Canada, Europe, and Russia, and will likely involve inhumane border control practices.

Imagining what this will actually mean at a national level is disheartening. For the United States, coastal cities in hurricane alley along the Gulf Coast will have to be abandoned, possibly as soon as the first half of the century, certainly by the end of the century. New Orleans will obviously be first, but Pascagoula and Bay St. Louis, Mississippi, and Houston and Beaumont, Texas, and other cities will be close behind. After the first couple of episodes of flooding and destructive winds, starting with Hurricanes Katrina and Rita in 2005, the cities will be partially rebuilt; the third major incident will make it clear that the risk of renewed destruction is too high to justify the cost of reconstruction. The abandonment of oil and natural gas production facilities in the Gulf region will push the United States into a severe recession or even depression, probably before the abrupt climate effects take hold in 2040. Mexico's economy will be devastated, which will increase illegal immigration into the United States.

Other major U.S. cities are likely to become uninhabitable after 2040, including New York City and Los Angeles, with a combined metropolitan population of nearly 33 million people. Resettling these populations will be a massive challenge that will preoccupy the United States, cause tremendous popular strife, and absorb all monies, including private donations, which would have previously gone to foreign aid. The United States, Canada, China, Europe, and Japan will have little choice but to become aggressively isolationist, with militarized borders. Given how dependent all these nations are on global trade, this will provoke a deep, persistent economic crisis. Standards of living across the United States will fall dramatically, which will provoke civil unrest across the country. The imposition of martial law is a possibility. Though the poor and middle class will be hit the hardest, no one will be immune. The fact that wealthier Americans will be able to manage the effects better, however, will certainly provoke resentment and probably violence and higher crime rates. Gated communities are likely to be commonplace. Finally, the level of popular anger toward the United States, as the leading historical contributor to climate change, will be astronomical. There will be an increase in asymmetric attacks on the American homeland.

India will cease to function as a nation, but before this occurs, Pakistan and Bangladesh will implode and help spur India's demise. This implosion will start with prolonged regional heat waves, which will quietly kill hundreds of thousands of people. It will not immediately be apparent that these are climate change casualties. Massive agricultural losses late in the first half of the century, along with the collapse of fisheries as a result of sea level rise, rising oceanic temperatures, and hypoxic conditions, will put the entire region into a food emergency. At first, the United States, Australia, China, New Zealand, and the Nordic nations will be able to coordinate emergency food aid and work with Indian scientists to introduce drought- and saltwater-resistant plant species. Millions of lives will be saved, and India will be stabilized for a time. But a succession of crippling droughts and heat waves in all of the donor nations and the inundation of several populous coastal cities will force these nations to concentrate on helping their own populations. The World Food Program and other international aid agencies will first have trouble operating in increasingly violent areas, and then, as donations dry up, will cease operations. Existing internal tensions in India will explode in the latter half of the century, as hundreds of millions of starving people begin to move, trying to find a way to survive. As noted above, a nuclear exchange between either the national governments or subnational groups in the region is possible and perhaps even likely.

By mid-century, communal genocide will rage unchecked in several African states, most notably Sudan and Senegal, where agriculture will completely collapse and the populations will depend on food imports. Both nations will be covered with ghost towns, where entire populations have either perished or fled; this will increasingly be true across Africa, South Asia, Central Asia, Central America, the Caribbean, South America, and Southeast Asia.

Europe will have the oddity of having to deal with far colder winters, given the collapse of the MOC, which will compromise agricultural productivity.

Major coastal cities, including London, will be struggling with inundation by mid-century. Throughout the century, as first thousands and then millions and then hundreds of millions of starving people begin flooding toward Europe, the EU will try to retreat behind high walls and naval blockades, a containment strategy that will be seen as morally indefensible and will provoke tremendous internal unrest and impoverishment, but also will be seen as a matter of survival. It is doubtful that the strategy will work, however, given the sheer numbers of migrants and the lack of resources to defend the borders.

China is likely to descend into anocracy, as the government struggles to deal with declining agricultural productivity, water scarcity, and the inundation of major coastal cities such as Shanghai. Internally displaced people, extremist and separatist movements, and refugees from other Asian and Eurasian areas will challenge Chinese stability. The nation may be able to hold itself together, but at great human cost. All of the impressive gains of the last twenty years will be lost in the coming century.

Throughout the century, there will be an urgency to achieving advances in agriculture, with the focus on drought- and flood-resistant species, as well as on varieties that can survive saltwater intrusion or can be cultivated in marginal soils. Even though a second "green revolution" takes place across the world, it neither stabilizes nor increases food supplies enough to keep up with demand and with shifting agricultural productivity patterns. The hardest hit nations are India, Senegal, Sudan, Pakistan, Bhutan, Myanmar, Thailand, Cambodia, Australia, Iran, Iraq, Syria, Yemen, Ethiopia, Morocco, Mali, Niger, Algeria, Guinea Bissau, Guinea, Sierra Leone, Liberia, Central African Republic, Congo, Gabon, Angola, Namibia, Botswana, South Africa, Zimbabwe, Zambia, Madagascar, Peru, Ecuador, Bolivia, Paraguay, Venezuela, Brazil, and the United States.[14]

Around the world, as human development indices fall, global impoverishment will become the norm. Anocracy will become common and autocracies will rise and fall as the level of violence and insecurity grows. The driving concern for most of the world population will be survival.

A Glimmer of Hope

The 2001 Broadway musical *Urinetown* projects a future of prolonged drought. In the play's dystopic vision, governance is taken over by an autocratic private company that controls sanitation and charges the public extortionate rates. Public urination is outlawed and punishable by death. The play treats hope and determination as dangerous delusions that are literally extinguished in the end.

Although this hundred-year climate change scenario is based on plausible scientific estimates, it is not meant to be a prediction, any more than *Urinetown* was meant to be a prediction. Such scenarios are an exercise in purposeful creativity, meant to help clarify the stakes for our actions today. There is one way, however, in which the catastrophic scenario is clearly unrealistic. Even in this extreme scenario, there would be countervailing trends and wild cards.

The strongest countervailing trends will be science and technology, and the social changes that come with them. As noted in the beginning of this chapter, innovations in computing, bioengineering, and nanotechnology will have a profound effect on societies around the world. These technologies may dramatically affect global energy consumption and cut greenhouse emissions, and they may give us more tools to adapt to the effects that can no longer be prevented. The ways in which information technologies and other advancements are knitting together global communities and empowering international organizations may strengthen our ability to anticipate and respond to climate change and shore up the benefits of globalization (such as international trade and collective action for humanitarian relief, peacekeeping, and other stabilizing purposes), when they come under pressure. And it is not outside the realm of possibility that we will find ways to counter the effects of climate change. Determination and necessity can be powerful drivers for innovation.

The greatest danger, in fact, may well be that the world will be too slow to realize that we are in the throes of catastrophic change. When people begin to die alone in greater numbers, from heatstroke and starvation, will we recognize these events as consequences of global warming? Will we add up the numbers and understand the scale of what is happening? Indeed, are we already missing a preponderance of evidence?

There are two other countervailing trends to take into account. One is the power of human relationships. A recent *Washington Post* profile of a Somali woman follows her flight from Mogadishu to a coastal town. She arrives with nothing but the worn-out clothes on her back and four sick children in tow. A distant cousin takes her in and shares what meager resources she has, diminishing her own chances of survival. For her, the ties of family, history, and empathy are stronger than the risk of starvation.[15] Indeed, the human capacity for hope and determination, and the efficacy of these qualities, should never be discounted, even in the most extreme of circumstances. Human beings do not only respond affirmatively to bright futures and opportunity; they also can respond with generosity and courage in adversity.

Finally, a grim countervailing trend might be a form of creative destruction, on a grand scale. If global catastrophe results from humanity's consumption patterns, the first thing that will happen is that our consumption patterns will change. With a global economic downturn and high mortality rates, people—even Americans, who consume a disproportionate share of world resources relative to the size of the population—will consume less, particularly less energy. Indeed, global energy supply systems are likely to become highly insecure long before the most catastrophic effects occur. We will either learn to consume less or we will find alternatives. Either way, scarcity may turn out to be the Earth's own mitigation strategy. The cost will be terrible for all people on the Earth, particularly for those who are already at a disadvantage, but that may ultimately be the price of survival for humanity.

Notes

1. Metropolitan Policy Program, Brookings Institution, and Greater New Orleans Community Data Center, "The New Orleans Index: Tracking Recovery of New Orleans and Metro Area," November 13, 2007 (www.gnocdc.org/KI/KatrinaIndex.pdf).

2. Jim Letten, United States Attorney, Eastern District of Louisiana, statement at hearings before Committee on the Judiciary, United States Senate, June 20, 2007.

3. H. G. Wells, *Anticipations of the Reaction of Mechanical and Scientific Progress Upon Human Life and Thought* (London: Chapman & Hall, 1902).

4. John Elfreth Watkins Jr., "What May Happen in the Next 100 Years," *Ladies Home Journal,* December 1900, as cited at www.paleofuture.com.

5. J. R. Minkel, "100 Year Forecast: New Climate Zones Humans Have Never Seen," *Scientific American,* March 26, 2007.

6. Jerome C. Glenn and Theodore J. Gordon, *2007 State of the Future* (Millennium Project, 2007; see www.millennium-project.org), p. 3.

7. Ray Kurzweil, *The Singularity Is Near: When Humans Transcend Biology* (New York: Penguin Books, 2005).

8. Please see chapter 3 of this volume for an explanation of the MOC.

9. Claude A. Piantadosi, *The Biology of Human Survival: Life and Death in Extreme Environments* (Oxford University Press, 2003), p. 75.

10. Ted Dagne, "Somalia: Current Conditions and Prospects for a Lasting Peace," Congressional Research Service Report for Congress, March 2007, p. 17.

11. UN Security Council, *Report of the Secretary-General on the Situation in Somalia,* Secretary General's Report S/2007/658 (New York: November 7, 2007).

12. Dan Smith and Janani Vivekananda, *A Climate of Conflict: The Links Between Climate Change, Peace, and War* (London: International Alert, 2007).

13. German Advisory Council on Global Change, *World in Transition: Climate Change as a Security Risk,* "Summary for Policy-Makers" (Berlin: May 2007), p. 6 (www.wbgu.de/wbgu_jg2007_kurz_engl.html).

14. William R. Cline, *Global Warming and Agriculture: Impact Estimates by Country* (Washington: Center for Global Development and Peterson Institute for International Economics, July 2007).

15. Stephanie McCrummen, "A Wealth of Kindness among Somalia's Poorest," *Washington Post,* December 10, 2007, p. A1.

seven

A Partnership Deal: Malevolent and Malignant Threats

R. JAMES WOOLSEY

> [There is] a tendency in our planning to confuse the unfamiliar with the improbable. The contingency we have not considered looks strange; what looks strange is therefore improbable; what seems improbable need not be considered seriously.
>
> —Thomas C. Schelling, foreword, *Pearl Harbor: Warning and Decision* (1962)

> Year after year the worriers and fretters would come to me with awful predictions of the outbreak of war. I denied it each time. I was only wrong twice.
>
> —Senior British intelligence official, retiring in 1950 after forty-seven years of service, quoted in Amory Lovins and Hunter Lovins, *Brittle Power: Energy Strategy for National Security*

The first two scenarios in this exercise dealt generally with climate change, the role of greenhouse gas emissions therein, and the regional consequences of smaller but substantial changes—up to a temperature rise of 2.6°C (4.7°F) and sea level rise of approximately half a meter (1.6 feet) in a thirty-year period. The third scenario discussed catastrophic change where aggregate global temperature increased by 5.6°C (10.1°F) by the end of the century, accompanied by a dramatic rise in global sea levels of 2 meters (6.6 feet) in the same time period. We might call climate change a "malignant," as distinct from a "malevolent," problem—a problem of the sort

Einstein once characterized as sophisticated (*raffiniert*) but, being derived from nature, not driven by an evil-intentioned (*boshaft*) adversary.

Sophisticated malignant problems can still be awesomely challenging. For example, because complex systems can magnify even minor disturbances in unpredictable ways—the so-called butterfly effect—a tree branch touching some power lines in Ohio during a storm can produce a grid collapse. In 2003 such a tree branch–power line connection deprived the northeastern United States and eastern Canada of electricity for some days. Similarly, our purchases today of gas-guzzling SUVs can contribute to sinking portions of Bangladesh and Florida beneath the waves some decades hence. With respect to climate change three factors should lead a prudent individual to consider such catastrophic change plausible: first, the possibility that some positive feedback loops could radically accelerate climate change well beyond what the climate models currently predict; second, the prospect of accelerated emissions of carbon dioxide (CO_2) in the near future due to substantial economic and population growth, particularly in developing countries such as China; and third, the interactive effects between these two phenomena and our increasingly integrated and fragile just-in-time—but certainly not just-in-case—globalized economy.

Exponential Change and Scenario Planning

The possibility of catastrophic exponential change necessitates a unique approach. This is because few human beings naturally think in terms of the possibility of the exponential changes. We humans generally have what the inventor and futurist Ray Kurzweil calls an "intuitive linear" view of phenomena rather than a "historical exponential" view. In *The Singularity Is Near,* he uses the example of a property owner with a pond who frequently cleans out small numbers of lily pads. Then, with the pads covering only 1 percent of the pond, the owner goes away, but he returns weeks later to find it covered with lily pads and the fish dead.[1] The owner, because the human mind thinks linearly, forgot that lily pads reproduce exponentially. When change is exponential we often have great difficulty comprehending it, whether it is manifested in lily pad growth or climatological tipping points. A related difficulty is that the adaptability of human society itself is difficult to predict in the presence of great and continuing catastrophe. The conflicts over land, migrating populations, or resources described elsewhere in this study might well be overshadowed in such a case by broader societal collapse.

Massively Destructive Terrorism

Another growing threat also holds out the possibility of massive damage and loss of life in this century: religiously rooted terrorism. The scope of death and destruction sought by the perpetrators of this sort of terrorism is also something most people find difficult to envision. This chapter later discusses terrorism (a "malevolent" rather than a "malignant" problem such as climate change) because of a somewhat surprising confluence: the aspects of our energy systems that help create the risk of climate change also create vulnerabilities that terrorists bent on massive destruction are likely to target. We need to be alert to the possibility that although our current circumstances are doubly dangerous, this confluence could give us an opportunity to design a set of changes in our energy systems that will help us deal with both problems.

Positive Feedback Loops and Tipping Points

The climate models agreed upon by the Intergovernmental Panel on Climate Change (IPCC) deal with some, but by no means all, of the warming effects of emissions that can occur as a result of positive feedback loops. This is because climatologists, as scientists, are given to producing testable hypotheses and there are often not enough data to satisfy that requirement for a number of the feedback loop issues. But a number of climatologists have nevertheless assessed the data and offered judgments about the importance of possible feedback effects, even in this century. NASA's James Hansen puts it succinctly: "I'm a modeler, too, but I rate data higher than models."[2] Positive feedback loops can relatively quickly accelerate climate change to the tipping point, at which it becomes impossible to reverse destructive trends, even with future reductions of greenhouse gas emissions from human activities. Several such positive feedback loops are conceivable in this century, such as the risk that freshwater from melting Greenland glaciers would slow the North Atlantic meridional overturning circulation, changing ocean currents and attenuating the Gulf Stream's ability to warm Europe.

Polar Regions

Tipping points at which there might be irreversible thawing of Arctic permafrost or the melting and breakup of the West Antarctic and the Greenland ice sheets have such stunning implications they deserve particular attention. Somewhere around a million square miles of northern tundra are underlain by frozen permafrost containing about 950 billion tons of carbon—more than currently resides in the atmosphere.[3] If the permafrost were to thaw, much of this carbon would quickly convert to methane gas. At about one million tons annually, the increase in atmospheric methane content is much smaller than the increase in CO_2 content, which weighs in at about 15 billion tons per year.[4] However, a ton of methane affects climate twenty-five times more powerfully than a ton of CO_2 over a 100-year time horizon.[5] As a result, it would take only 600 million tons of methane to equal the global warming effect of 15 billion tons of CO_2. If this seems like an implausibly large increase in methane emissions, consider that it equates to only one-half of one-tenth of 1 percent of the organic carbon currently preserved in the permafrost (not to mention much larger amounts of frozen methane stored in shallow marine sediments). Therefore, if the permafrost begins to thaw quickly due to the initial linear warming trend we are experiencing today, the climate impact of methane emissions could come to rival that of CO_2 in future decades. Consequent accelerated warming and faster thaw leading to more methane emissions could produce a tipping point beyond which humans no longer control the addition of excess greenhouse gases to the atmosphere, and no options remain under our control for cooling the climate. We don't know the exact point at which this vicious circle would begin, but there are some indications that a substantial permafrost thaw is already under way.[6]

Because of methane's potency its release could provide a substantial short-term kick to climate change. Such release over a few decades could raise worldwide temperatures by 5 to 6°C (9 to 10.8°F) or more,[7] to the approximate level of temperature increase posited for the third scenario in this study. Another potential feedback loop lurks in the prospect of melting—and sliding—ice sheets in Greenland and West Antarctica. Around 125,000 years ago, at the warmest point between the last two ice ages, global sea level was 4 to 6 meters (about 13 to 20 feet) higher than it is today and global temperature was only about 1°C (1.8°F) higher.[8] Being warmer than Antarctica, Greenland probably provided the initial slug of meltwater to the ocean. However, much of the ice of western Antarctica rests on bedrock far below sea level, making it less stable as sea level rises.[9] When the ice sheet is lubricated by melting where it is grounded, it begins to float and can cause coastal ice shelves to shatter, increasing the rate of ice stream flow into the ocean (*ice stream* is a region of an ice sheet that moves significantly faster than the surrounding ice).[10] As a result of this action, the West Antarctic Ice Sheet contributed perhaps 2 meters (6 or 7 feet) of the additional sea level 125,000 years ago.

With just 1°C (1.8°F) of warming, therefore, we may be locked into about 4 to 6 meters (13 to 20 feet) of sea level rise.[11] James Hansen points out that it is not irrational to worry about reaching this tipping point in this century. This study's catastrophic scenario assumes 5 to 6°C (9 to 10.8°F) of warming, which is significantly warmer than conditions 3 million years ago, before the ice ages. At that time, the Earth was 2 to 3°C (3.6 to 5.4°F) warmer and sea level was about 25 meters (82 feet) higher than today.[12] Although the time required for that much sea level rise to occur is probably more than 1,000 years, the third scenario, with 2 meters (6.6 feet) of sea level rise by the end of this century, appears quite plausible.[13]

Economic Development

Robert Zubrin, the author of *Energy Victory: Winning the War on Terror by Breaking Free of Oil,* who is something of a climate change skeptic, suggests a simple thought experiment to illustrate the power of economic growth to affect climate change—a process that could create a climatic tipping point sooner rather than later. The world today has achieved an average GDP per capita comparable to U.S. GDP per capita at the beginning of the twentieth century (about $5,000 in today's dollars).[14] In the twentieth century, world population quadrupled and world economic growth averaged 3.6 percent annually.[15] Even if we assume slower population growth, say a doubling of world population in the twenty-first century, and also a lower growth rate of 2.4 percent—the latter producing a fivefold increase in GDP per capita—unless fuel use per unit of GDP changes substantially, we would see a tenfold increase in CO_2 emissions by century's end. This prospect leads even a climate change skeptic such as Zubrin to imagine an extraordinary scenario in which presumably all known and some unknown feedback loops become activated and thus it "only tak[es] a few decades to reach Eocene carbon dioxide atmospheric concentrations of 2,000 ppm"—and certain catastrophe.[16]

To take only one example of the impact of vigorous economic development on CO_2 emissions, China is building approximately one large coal-fired power plant per week for the foreseeable future. Rapidly growing developing countries are expected to account for an overwhelming 85 percent of energy-demand growth between 2008 and 2020. China alone represents a third of total growth.[17]

Sea Level Rise and Challenges to Existing Infrastructure

The *2007 IPCC Working Group I Contribution to the Fourth Assessment Report* points out that the prospect of climate change and sea level rise coming to a tipping point is particularly troubling because once such a point has been passed, sea level rise will probably continue for centuries.[18] For this reason,

James Hansen considers sea level rise as "*the* big global issue" that will transcend all others in the coming century.[19] Even if the East Antarctic Ice Sheet is not destabilized, the steady melting of the Greenland Ice Sheet together with the perhaps sudden melting of the West Antarctic Ice Sheet holds the potential for some 12 meters (40 feet) of sea level rise.[20] The melting of the East Antarctic shelf would add approximately 25 meters (80 feet); this would mark, in the Antarctic research scholar Peter Barrett's words, "the end of civilization as we know it."[21] Even without a melting of the East Antarctic shelf, civilization would be experiencing an inexorable encroachment of seawater over decades and centuries.

Moreover, humanity would have to face the coastal inundation and related destruction while dealing with substantial disruption of agriculture and food supplies, and resulting economic deprivation, due to changing availability of water—some places more arid, some wetter—and a much smaller percentage of available water would be fresh.

Coastal Regions

The catastrophic scenario outlined in chapter 6 listed among the regions in the developed world facing the likely prospect of inundation by the end of the century: major portions of cities and wide regions of the U.S. coast from South Texas to West Florida and from East Florida to New York; extensive areas bordering the Chesapeake Bay and most of South Florida and eastern North Carolina; the lower Hudson River Valley; huge shares of the coasts of San Francisco Bay; much of Sydney and all of Darwin, Australia; a large share of Japanese ports; Venice and a major share of coastal Tuscany; the majority of the Netherlands; much of Dublin; a major share of Copenhagen; and the Thames River Valley and the eastern and southern coasts of England.[22] Storm surges would affect people much farther inland and on more elevated coastlines.

Even without considering storm surge, sea level rise in the range of 2 meters (6.6 feet) in this century could have a potentially catastrophic effect on a number of developing countries. According to a February 2007 World Bank policy research working paper, these include particularly Egypt, Vietnam, and the Bahamas and a number of other island nations. It could also have "very large" effects on a number of other states, including China and India. Considering all factors—land area, urban area, population, and so forth—the most affected countries, in addition to those just cited, would be Guyana, Surinam, and Mauritania. Substantial impacts would also occur in Gambia, Liberia, Senegal, Guinea, Thailand, Burma, Indonesia, Taiwan, Bangladesh, and Sri Lanka.

A 2-meter (6.6-foot) rise in sea levels—together with changed climate, agricultural disruptions and famines, the spread of disease, water scarcity, and severe storm damage—will not occur in a world that is otherwise sustainable and resilient. Many areas are already destabilized. In the Philippines, for example, sea level rise would add to a problem already created by excessive groundwater extraction, which is lowering the land annually by fractions of an inch in some spots to more than a tenth of a meter (3 or 4 inches) annually.[23] The Mississippi Delta has a similar problem with land subsidence. Some of the land south of New Orleans will likely lose about 1 meter (3 feet) of elevation by the end of this century as a result of subsidence.[24] Thus, about 6 feet (about 2 meters) of sea level rise by the end of the century may well be additive to the substantial lowering of land levels in some areas by such groundwater extraction. And the concentration of population in low-lying areas of course exacerbates the effect of these changes.

Meltwater runoff from mountain glaciers also supplies agricultural and drinking water as well as electricity from hydropower. More than 100 million people in South America and 1 billion to 2 billion in Asia rely on glacial runoff for all or part of their freshwater supply. As these glaciers shrink and produce less meltwater they will contribute substantially to the need to emigrate in search of water and arable land. The relevant glaciers are retreating rapidly and some are already virtually gone. This problem is likely to peak within mere decades.[25]

Potential National Security Consequences of Climate Change

In a world that sees a 2-meter (6.6-foot) sea level rise with continued flooding ahead, it will take extraordinary effort for the United States, or indeed any country, to look beyond its own salvation. All of the ways in which human beings have responded to natural disasters in the past, which John R. McNeill describes in chapter 2, could come together in one conflagration: rage at government's inability to deal with the abrupt and unpredictable crises; religious fervor and perhaps even a dramatic rise in millennial end-of-days cults; hostility and violence toward migrants and minority groups, at a time of demographic change and increased global migration; and intra- and interstate conflict over resources, particularly food and freshwater.

Altruism and generosity would likely be blunted. In a world with millions of people migrating out of coastal areas and ports across the globe, it will be extremely difficult, perhaps impossible, for the United States to replicate the kind of professional and generous assistance provided to Indonesia following the 2004 tsunami. Even overseas deployments in response to clear military needs may prove very difficult. Nuclear-powered aircraft carriers and submarines might be able to deploy, but aviation fuel or fuel for destroyers and other non-nuclear ships could be unobtainable. Overseas air bases would doubtless be tangled in climatic chaos, and aircraft fuel availability overseas highly uncertain. Further, the Navy is likely to be principally involved in finding ways to base, operate, overhaul, and construct ships, as many ports and harbors south of New York on the East Coast and overseas disappear or become usable only with massive expenditures for protection from the rise in sea levels. Civilians will likely flee coastal regions around the world, including in the United States. The U.S. military's worldwide reach could be reduced substantially by logistics and the demand of missions near our shores.

Population Changes and Migrations

If Americans have difficulty reaching a reasonable compromise on immigration legislation today, consider what such a debate would be like if we were struggling to resettle millions of our own citizens—driven by high water from the Gulf of Mexico, South Florida, and much of the East Coast reaching nearly to New England—even as we witnessed the northward migration of large populations from Latin America and the Caribbean. Such migration will likely be one of the Western Hemisphere's early social consequences of climate change and sea level rise of these orders of magnitude. Issues deriving from inundation of a large portion of our own territory, together with migration toward our borders by millions of our hungry and thirsty southern neighbors, are likely to dominate U.S. security and humanitarian concerns. Globally as well, populations will migrate from increasingly hot and dry climates to more temperate ones.

On the other hand, extrapolating from current demographic trends, we estimate that there will be fewer than 100 million Russians by 2050, nearly a third of whom will be Muslims. Even a Europe made colder by the degrading of the Gulf Stream may experience substantially increased levels of immigration from south of the Mediterranean, both from sub-Saharan Africa and from the Arab world. Many of Europe's Muslim minorities, including Russia's, are not well assimilated today, and the stress of major climate change and sea level rise may well foster social disruption and radicalization. Russia and Europe may be destabilized, shifting the global balance of power.

Northern Eurasian stability could also be substantially affected by China's need to resettle many tens, even hundreds, of millions from its flooding southern coasts. China has never recognized many of the Czarist appropriations of north Central Asia, and Siberia may be more agriculturally productive after a 5 to 6°C (9 to 10.8°F) rise in temperatures, adding another attractive feature to a region rich in oil, gas, and minerals. A small Russian population might have substantial difficulty preventing China from asserting control over much of Siberia and the Russian Far East. The probability of conflict between two destabilized nuclear powers would seem high.

Energy Infrastructure

Interactions between climate change and the existing infrastructure could create major failures in the systems that support modern civilization. All other systems—from operating telecommunications to distributing food, pumping water, and more—depend on energy. Yet energy systems themselves are vulnerable. Hydroelectric electricity generation may be substantially affected by reduced glacial runoff or by upstream nations diverting rivers in some parts of the world. Nuclear power plant cooling may be limited by reduced water availability. Increased numbers and intensity of storms could interfere with long-distance electricity transmission, already heavily stressed in the United States and elsewhere.

Sea level rise and chaotic weather patterns may interfere with oil production in a number of locations, particularly from sea-based platforms and in parts of the Middle East, and with the operation of large oil tankers. Many U.S. oil refineries are in the Gulf Coast region and thus more vulnerable to disruption by storms than if they were located elsewhere. Hurricane Katrina came very close to shutting down the Colonial Pipeline, the major link from the Gulf Coast to the Eastern Seaboard. In short, the pressures on U.S. society and the world would be significant, and the international community's ability to relieve those pressures seriously compromised. The abrupt, unpredictable, and relentless nature of the challenges will likely produce a pervasive sense of hopelessness.

A Malevolent Threat: Mass Terrorism

Our society, our way of life, and our liberty face serious current challenges beyond the infrastructure fragility exacerbated by climate change. The most salient is attack by terrorist groups or an enemy state, or a combination thereof, aimed at massive damage and massive casualties. These are not unintentional "malignant" results of our habitual behavior but are rather "malevolent" and planned carefully by those who want to do far more than many terrorist groups in the past: namely, to destroy our entire civilization and way of life.

Oil presents a panoply of opportunities for highly destructive terrorism. Our transportation is fueled over 96 percent by petroleum products. Consequently oil has a transportation monopoly in much the same way that, until around the end of the nineteenth century, salt had a monopoly on the preservation of meat. Oil's monopoly creates a litany of vulnerabilities for our society.

Since around two-thirds of the world's proven reserves of conventionally produced oil are in the Persian Gulf region, together with much of oil's international infrastructure, the world's supplies are vulnerable to terrorist attacks such as two already attempted by al Qaeda in Saudi Arabia and emphasized in al Qaeda's doctrine. Some oil states' governments (Iran) are quite hostile today; others (Saudi Arabia) could become so with a change of ruler. A nuclear arms race appears to be beginning between Iran and six Sunni states that have announced nuclear programs "for electricity generation." The United States borrows more than a billion dollars a day at today's prices to import oil, substantially weakening the dollar. The Wahhabi sect of Saudi Arabia profits massively from oil income and, according to Lawrence Wright in *The Looming Tower: Al-Qaeda and the Road to 9/11,* covers "90 percent of the expenses of the entire faith, overriding other traditions of Islam."[26] Wahhabi teachings are murderous with respect to Shi'ite Muslims, Jews, homosexuals, and apostates; are hideously repressive of women; and are mirrored by the views of al Qaeda and similar groups except with respect to their allegiance to the Saudi state. And finally, as Bernard Lewis puts it, "There should be no taxation without representation but it should also be noted that there is no representation without taxation." Extremely wealthy oil-exporting states are thus often dictatorships and autocratic kingdoms without institutions that check and balance the ruler.

The other major energy sector of our economy, electricity generation and distribution, is also highly vulnerable to attack by terrorists and rogue states. In 2002 the National Research Council published its report on the use of science and technology to combat terrorism. It stated: "The most insidious and economically harmful attack would be one that exploits the vulnerabilities of an integrated electric power grid. 'A chain is only as strong as its weakest link' applies here. Simultaneous attacks on a few critical components of the grid could result in a widespread and extended blackout. Conceivably, they could also cause the grid to collapse, with cascading failures in equipment far from the attacks, leading to an even larger long-term blackout."[27]

As of 2008 very little has been done to implement the council's seventeen detailed recommendations to deal with this, particularly with regard to improving the security of, or even stockpiling spares for, the large transformers at grid substations or effectively protecting the grid's Supervisory Control and Data Acquisition (SCADA) control systems from destructive hacking. Additionally, the electricity grid has a major vulnerability to an electromagnetic pulse (EMP). In 1962 both Soviet and American atmospheric nuclear tests revealed a troubling phenomenon: three types of electromagnetic pulses generated at high altitude by nuclear detonations could seriously damage or destroy electronic and electrical systems at as much as 1,610 kilometers (1,000 miles) from the blast. The 2004 report of the U.S. Electromagnetic Pulse Commission pointed out that the detonation of a single nuclear warhead between 40 and 400 kilometers (25 and 250 miles) above the Earth could cause "unprecedented cascading failures of our major infrastructures," primarily "through our electric power infrastructure" crippling "telecommunications . . . the financial system . . . means of getting food, water, and medical care to the citizenry . . . trade . . . and production of goods and services." The commission noted that states such as North Korea and Iran, possibly working through terrorist groups, might not be deterred from attack (say using a relatively small ship carrying a simple SCUD missile) in the same way as were our adversaries in the cold war.[28]

The commission concluded that detonation of a single nuclear warhead at these altitudes could "encompass and degrade at least 70 percent of the Nation's electrical service, all in one instant." It also notes that, as a result of fire safety and environmental concerns, locally stored fuel for emergency power supplies such as diesel for generators is often limited to about a seventy-two hours' supply.[29] Food available in supermarkets generally supplies about one to three days of requirements for customers, and regional food warehouses usually stock enough for a multicounty area to last about one month.[30]

Toward a Partnership to Deal with Both Malignant and Malevolent Threats

These malignant and malevolent risks seem to stem from very different causes—and different kinds of people, with different backgrounds, tend to look at them separately. This cultural separation—analogous in some ways to C. P. Snow's famous description some decades ago of the intellectual world's division into the two cultures of literature and science—hinders cooperative action. For the issues at hand, let's call this a division between the tree-hugger culture, focused on carbon, and the hawk culture, focused on terrorism.

Both the malignant and malevolent problems described here are extraordinarily grave, and much too urgent to await a lengthy debate between the two cultures about how intensely we should believe that each risk will become manifest. This is especially true because, as suggested below, the steps needed to contend successfully against both types of problems appear to have a great deal in common, at least in the important field of energy.

A hawk who is steeped in the history of the Muslim Brotherhood but has no time for the history of glaciers need not be required to pledge his belief that climate change will hit a certain degree by a certain date. Scientific theories, Karl Popper taught us, must always be held tentatively; they are productive precisely to the degree that they offer an invitation to be *dis*proved. Even as society used Newton's theories for centuries, the path of human progress was to give others a chance to create theories that would replace his. Eventually Einstein's did.

Nevertheless, we should argue to our hawk that as a matter of judgment, not certainty, there is sufficient evidence of developing climate change that he or she should take the issue seriously. Further, if we consider together plausible climatic tipping points and the increased emissions from world economic development, there is a risk that such change could become cataclysmic. Thus, the only responsible course of action is to begin now to deal with the problem as sensibly and affordably as we can.

We should say something similar to a tree hugger who is quite attentive to possible change in the North Atlantic meridional overturning circulation but who believes that to deal with terrorism now and for the foreseeable future we need only enforce the criminal law—and that a rogue state or terrorist EMP attack on the United States must be someone's idea of a film plot for the PG-13 market. The tree hugger's blind spot is precisely where the hawk's eyes are trained, and vice versa. But our tree hugger needs to remember that fanatic enemies with access to destructive technology have already wreaked mass death on modern societies. The tree hugger needs to keep an open mind, remember the Nazis, and recognize that evil exists, and happens.

As a thought experiment we might try inviting a tree hugger, someone strongly committed to reducing the risk of climate change, to address a major malignant issue by producing a short list of policies that could soon lead to substantial reductions of emissions. We will ask the tree hugger to focus on the ways in which we generate electricity, fuel transportation, power industry, and operate buildings, leaving such topics as preventing deforestation and promoting proper agricultural practices until later. We want him to focus on energy because we are going to submit his list to someone else for comment—a hawk who is heavily focused on energy security—to see if there is anything on which they can agree.

For our tree hugger we decide to summon the shade of John Muir, the father of our national parks system and the first president of the Sierra Club, and for our hawk, the shade of George S. Patton, commander of the Third Army in World War II. They eye each other warily, but agree to undertake our project.

After sitting and pondering thoughtfully for a time under some redwoods, Muir submits a list of nine proposals for Patton's consideration:

1. Begin with improving the energy efficiency of buildings.

Muir notes that Wal-Mart is finding that with such simple steps as painting its store roofs white and adding skylights, the company is getting 20 percent improvement in energy efficiency today and expects 25 to 30 percent improvements by 2009. And Muir has seen a recent McKinsey & Company report that says that merely by using existing technologies (where there is an internal rate of return of 10 percent or more) we can reduce world energy demand by 125 to 145 QBTUs (quadrillion British thermal units) by 2020, 20 to 24 percent of end-use demand. The vast majority of this, the report says, would be in buildings of all sorts, including industrial facilities, and would contribute up to half the greenhouse gas emission abatement needed to cap the long-term concentration of greenhouse gases in the atmosphere at 450 to 550 ppm.[31] Muir knows that the Rocky Mountain Institute's thorough work shows even more opportunity for energy savings from reduced energy use in buildings.[32]

"I'm completely with you on this one," says Patton. "Less need for energy, less need to add generating capacity and transmission lines to the grid. Every day, the grid reminds me more and more of the Maginot Line, just sitting there vulnerable to being taken out by creative tactics—the less we need it the better. And I like the fact that this efficiency stuff makes money for the folks who implement it rather than costing something."

2. Radically increase the use of combined heat and power (CHP).

His second item, Muir says, could be implemented relatively quickly and would let us get dual use from energy instead of wasting a lot of the heat our industry produces by just venting it into the atmosphere. About a third of Denmark's electricity, for example, comes from CHP. Only about 8 percent of

U.S. electricity comes from CHP, but the problem—like building efficiency—is not that we don't have the technology. Rather, Muir says, our commitment to wasting heat is determined by culture and regulations. Much of the reason CHP struggles in the United States is because of the opposition of state public utility commissions (PUCs). Certain steps are needed to ensure safety, Muir concedes, but the Danes have figured this out and completely changed their system in just twenty years. To do what they've done we just need to change most states' PUC policies. CHP generally has the effect of generating electricity and heat closer to where they are used, in relatively small facilities, Muir notes.

"Go, Danes!" says Patton. "You know, John," he continues, "I admit I was pretty skeptical when I agreed to do this with you, but I've gotta admit I'm learning some things and I like this one, too. Just using energy we're already producing—makes all the sense in the world. And it looks like each of these two ideas of yours reduces the need for new centralized power generation plants as well as new long-distance transmission lines. Relying on smaller, more distributed, production should improve resilience against terrorist attack. Keep 'em coming."

3. Create strong long-term incentives for small-scale (single-building-based) distributed generation of electricity and heating and cooling.

Forty out of fifty states, Muir says, now have "net metering" laws that in principle make it possible for those who have generating capacity—say roof-top solar photovoltaic systems—to sell some home-generated electric power back to the grid. But in practical terms, state laws and regulations leave a lot to be desired in making this work. The cost of home-generated power is about to decline sharply, says Muir. As thin-film and nano-solar technologies come on the market at costs substantially below those of today's silicon cells, and as solar collectors are integrated into building materials such as shingles, these technologies can begin to have a substantial effect on the need for central power generation. Small-scale wind turbines, operating at lower wind speeds than the large wind turbines, are beginning to come into the single-building market as well. Distributed solar and wind technologies complement one another, since generally the sun shines at a different time of day than the wind blows, and increased use of both can be facilitated by storing electricity in improving batteries. Shallow (heat pump) geothermal is showing promise for heating and cooling of individual buildings; together with distributed solar and wind it may be able to satisfy a very substantial share of individual building energy needs. Distributed generation will be renewable and hence not carbon-emitting, Muir notes: a coal-fired power plant will not fit on a roof.

"John," says Patton, "anyone who has ever been in combat knows that you need flexibility and initiative at the small-unit level because the unexpected always happens, and if your small units are good you can adapt faster. I've always said, "Small had damned well better be beautiful." You have to be able to put maximum reliance on your platoon leaders and sergeants—that's how I was able to relieve Bastogne so fast. You're making me see that the same logic applies to having an energy system that's resilient against terrorist and EMP attack. Damn, are you sure you don't have a military background?"

4. Follow California's lead and decouple sales from earnings for electric utilities to encourage conservation and grid modernization.

This is a big one, says Muir. California, he notes, initiated this simple step some twenty years ago; there, and (very recently) in several other states, utilities' earnings are based on their investment, not their sales of electricity. But in the other forty-plus states, utilities must sell more electricity in order to earn more for their shareholders. It doesn't matter if it's used wastefully—the incentive systems established by forty-some PUCs don't deter waste. In California and the other few states, though, if a utility invests in making the grid "smarter," say, to help consumers conserve electricity, it earns more for its shareholders. The effect of decoupling sales from earnings is dramatic: over the last twenty years, electricity use per capita in California has stayed flat, while that of the rest of the country has increased 60 percent. Major double-digit improvements in energy efficiency are possible if the other approximately forty PUCs would just admit that what a few states have done is problem-solving and that their own current policies are problem-creating.

"Sounds great," says Patton. "I know California screwed up on the Enron thing a while back—hell, everybody screws up sometimes—even I did once. But the Californians sure have this decoupling right. Say, who writes those other forty PUCs' fitness reports? Why don't their superior officers just relieve them of command and put somebody in charge who's willing to learn from what the California folks have done?"

5. Give steady and long-term encouragement to the deployment of renewable electricity generation for the grid from wind, solar, hydro, and geothermal.

Muir tells Patton that many incentives such as tax credits for such deployment have been periodically interrupted, delaying, for example, production of wind turbines and slowing the introduction of these technologies.

"Well," says Patton, "if we have to add to the grid I suppose these are okay. The grid will be around for a long time, so we have to improve its resilience by stockpiling transformers and defending better against cyber attacks in any case. But even if we improve its defenses and make it cleaner, increasing our reliance on a Maginot Line is not my favorite way to go. I liked your efficiency and CHP and rooftop ideas better, but I guess I can go along with these—I like the fact that at least some of them probably won't be too large and can be distributed to some extent. Also, power plants using sun, wind, hydro and geothermal aren't vulnerable to terrorist interruption of their fuel supplies."

6. Vigorously develop carbon capture and storage (CCS) for coal-fired power plants.
Muir points out that this may well rely on the already-developed technology of integrated gasification combined cycle (IGCC) plants, which facilitates CO_2 capture. The hard part is sequestering the CO_2 permanently where it will not leak into the atmosphere. The CO_2 gas may be pumped into existing oil and gas wells to enhance recovery from them. Pumping it into saltwater aquifers deep beneath the earth also shows promise for long-term sequestration.

Again, Patton is only lukewarm. "Adding to the grid just gives the terrorists eyeing our transformers and the crazy guys with EMP attack plans a bigger target," he says. "But if we can't get all the power we need by implementing your ideas about reducing demand and increasing distributed generation, then I'm okay with this CCS stuff, but reluctantly."

7. Provide tax incentives for the purchase of plug-in hybrid gasoline-electric vehicles (PHEVs). Now for transportation, Muir says. GM has announced the production of the Chevrolet Volt plug-in hybrid (PHEV) in 2010 (they call it an "electric vehicle with range extension"); Toyota's Prius was designed originally with an all-electric mode for driving, so it is well on the way to being a plug-in once a battery more capable than that in the current Prius is supplied. Other manufacturers are gearing up to produce plug-ins as well. There are dozens of hybrid vehicles, principally Priuses, that their owners have converted into PHEVs using currently available batteries. A PHEV that is plugged into a standard 120-volt socket in a garage overnight can be driven 32 to 65 kilometers (20 to 40 miles) the next day on this charge. Once it reaches the end of the electricity supplied in its overnight charge it becomes an ordinary hybrid, using both gasoline and electricity until it can be charged again. These vehicles seem to be getting over 160 kpg (100 mpg) once their initial all-electric driving is factored in. (Muir suggests to Patton he take a look at the websites pluginamerica.com and calcars.org.

The average U.S. light vehicle is driven just over thirty miles a day, Muir adds. It is clear that, in addition to providing consumers the ability to drive for some tens of miles a day on inexpensive off-peak overnight electricity at a fraction of the cost of driving on gasoline, moving from a standard internal-combustion-engine vehicle to a PHEV reduces greenhouse gas emissions substantially. A recent Pacific Northwest National Laboratory study has estimated that if 73 percent of the current U.S. fleet of light-duty vehicles were converted to PHEVs that were able to drive just over thirty miles all-electrically and were charged during off-peak hours, no new power plants would be needed. Moreover this would displace 6.5 million barrels of oil equivalent per day, or approximately 52 percent of the nation's oil imports. The average reduction nationally of greenhouse gases would be in the range of 27 percent per car, more in states using little coal to produce electricity, around zero in heavy coal-using states.[33] And over time cleaning up the grid also cleans up PHEV emissions: as electricity production is modified—say, via renewables or coal with carbon capture and sequestration—CO_2 emissions are further reduced.

Finally, PHEVs can replace certain "ancillary services" that cost about $12 billion annually, such as fossil fuel purchases to stabilize and regulate the grid's operations and "spinning" reserves to deal with power outages. Keeping just a small number of PHEVs plugged into the grid after they are charged creates vehicle-to-grid (V2G) connections that replace fuel-consuming functions.[34] This can mean a lot less use of fossil fuel and also substantial payments back to plug-in hybrid owners. One Federal Energy Regulatory Commission member even calls plug-ins "cash-back hybrids." Grid modernization can help implement such major innovations.

"John, now you're talking again," says Patton. "Electricity (and plug-ins) can do to oil what electricity (and refrigeration) did to salt around the time I was born—destroy the damned stuff as a strategic commodity. Salt used to be a really big deal because it was the only way to preserve meat. People even fought wars over it. But now nobody gives a damn what country has salt mines. Since around the time I commanded the Third Army, maybe before, the number one strategic commodity has been oil. It sure was in the war. If old Tooey Spaatz, God bless him, hadn't persuaded FDR to let him hit Ploesti and Leuna and take out the Germans' fuel, they would have had enough for the Panzers to get to Antwerp and the Battle of the Bulge could have gone the other way."

Patton shakes his head sadly. "You know, John, there are some jaspers at the Council on Foreign something-or-other in New York who say we're doing a 'disservice to the nation' by trying to get the country away from oil dependence. Do they think it's a 'service' to make it easier for some other country to have the leverage over us that we had over the Germans in the war? Those guys would probably also tell drunkards to make sure they have a glass or two of red wine every day for their health—not crazy in the abstract, but sure as hell not the message a guy in his cups needs to hear. But you're telling those council guys to get with the program and help get us off oil fast—John, you're my man."

8. Mandate a rapid transition to flexible fuel vehicles (FFVs).
Muir says this is simple, and would mean that both U.S.-produced vehicles and imports could use at least gasoline, ethanol (particularly cellulosic), butanol, and methanol in any mixture. This would create a market for renewable fuels by removing a needless barrier, Muir points out. He adds that using such fuels can substantially reduce greenhouse gas emissions, especially when the feedstocks are biomass and waste. The cost is modest—around $100 per vehicle or less. Between 2002 and 2005, Brazil moved from 5 percent to 75 percent of their new vehicles' being FFVs. Incentives such as tax credits should be provided promptly to encourage pumps for these fuels to be installed at stations.

"Hey, John," Patton booms. "I'm fine with markets and cap-and-trade and all that, but sometimes ya gotta just tell people to, damn it, *do* it. I got no problem with mandates—hell, if you gotta move fast and it's important, I absolutely *love* 'em. We did it for cars with seat belts and air bags because people's lives were at stake. Well, they're at stake because of oil dependence too. Getting away from that dependence is a matter of national security. Somebody just needs to show as much gumption as the Brazilians and issue a damned *order* about obvious stuff like this."

9. Provide incentives for the production of renewable fuels and specialty chemicals from cellulosic biomass; give special attention to the desirability of using waste products as a feedstock, particularly where methane is thereby reduced.
Muir points out that we should be moving away from hydrocarbons and toward carbohydrates generally as feedstocks for liquid fuels, electricity generation, and chemical production. But he is especially worried about a number of wastes producing methane if left in their natural state because of the latter's potency as a greenhouse gas—more than twenty times that of CO_2.

"Fine with me, John," says Patton. "Let's clean stuff up while we get off oil—a threefer: helps thwart the terrorists, reduces that carbon you're so worried about, and things smell better. I'm gonna start calling you 'God's janitor.' Basically you're nine for nine. Pretty interesting—we keep getting to the same place as long as we don't have to agree with one another's reasons for going there. Who'da thought it?"

"But there are three things you didn't mention," he adds: "Nuclear power, hydrogen, and coal-to-liquid transportation fuels. I've seen a lot of guys lobbying lately on all three of those—must be some money behind 'em. What do you think?"

Patton and Muir talk for a while and agree that nuclear power plants may be an acceptable last resort if we have to add generating capacity in the United States. Muir winces at the prospect, but in spite of the waste storage problem he's always been worried about, he's come reluctantly to support nuclear in some cases because of nuclear plants' lack of carbon emissions. Patton has a nagging problem with terrorist threats to power plants, but agrees that it would be very hard to cause a core meltdown. The two agree we should definitely oppose spreading nuclear energy around the world to new countries, since with today's treaties and inspections it's impossible in practical terms to stop countries from using their nuclear "electricity" programs as a way to get into the nuclear weapons business.

The hydrogen discussion just takes a few seconds. Both see some uses for hydrogen, but when they start talking about driving the "hydrogen highway" in family cars with hydrogen fuel cells and hydrogen storage and pumps at neighborhood filling stations, they shake their heads, amazed at the cost—especially, they chuckle, since the only infrastructure fueling cost you need for plug-in hybrids is an extension cord for each car-driving household.

Coal-to-liquids (CTL) is their only area of disagreement. Muir hates the carbon it would produce; Patton likes the way it undermines oil. As they finish their discussion, Patton puts a hand on Muir's shoulder and says, "John, tell you what I'll do. Even though CTL plants would use American coal, which I like, some plants might need a big infrastructure that could be vulnerable to terrorists, which I don't like. I'm happy with your transportation ideas because they move us toward small local plants and distributed production of fuel, whether electricity or liquid—nicely resilient. How about this: unless they figure out how to sequester enough of the carbon from CTL to satisfy you, I won't drop this option but I'll move it down to the bottom of my list—but in exchange I'd like a little help from you on another matter: I think the Army needs at least two to three more armored divisions. What do you say?"

"George," laughs Muir, "You're a piece of work. I might be able to talk myself into rolling over for one or two of those things, but, if I do, for each one I support I'm going to need your backing for at least one new national park."

"John," says Patton, "I like your style. Say, can you hunt in those places?"

"George," gasps Muir, "you are absolutely imposs—"

Patton grins. "Just pullin' your chain."

As they stroll off together into the evening haze, Patton's ghost begins slightly to resemble Humphrey Bogart, and Muir's, Claude Raines. Patton grins and says, "Y'know, Johnny, this could be the start of a beautiful friendship."

Notes

Epigraphs: Roberta Wohlstetter, *Pearl Harbor: Warning and Decision* (Stanford University Press, 1962); Amory Lovins and Hunter Lovins, *Brittle Power: Energy Strategy for National Security* (Andover, Mass.: Brick House, 1982).

1. Ray Kurzweil, *The Singularity Is Near* (New York: Viking, 2005), p. 8.
2. Fred Pearce, *With Speed and Violence: Why Scientists Fear Tipping Points in Climate Change* (Boston: Beacon Press, 2007): 58.
3. S. A. Zimov, E. A. G. Schuur, and F. S. Chapin III, "Permafrost and the Global Carbon Budget," *Science* 312 (2006): 1612–13.
4. L. Denman, G. Brasseur, A. Chidthaisong, and others, "Couplings between Changes in the Climate System and Biogeochemistry," in *Climate Change 2007: The Physical Science Basis; Contribution of Working Group I to the Fourth Assessment Report of the Intergovernmental Panel on Climate Change,* edited by Susan Solomon and others (Cambridge University Press, 2007).
5. P. Forster, V. Ramaswamy, P. Artaxo, and others, "Changes in Atmospheric Constituents and in Radiative Forcing," in *Climate Change 2007: The Physical Science Basis.*
6. Pearce, *With Speed and Violence,* pp. 77–85.
7. Ibid., p. 85 (citing L. Smith).
8. J. T. Overpeck and others, "Paleoclimatic Evidence for Future Ice-Sheet Instability and Rapid Sea-Level Rise," *Science* 311 (2006): 1747–50.
9. Joseph Romm, *Hell and High Water: Global Warming: The Solution and the Politics—and What We Should Do* (New York: HarperCollins, 2007), p. 86.
10. Ibid., p. 86; Pearce, *With Speed and Violence,* p. 58.
11. Overpeck and others, "Paleoclimatic Evidence for Future Ice-Sheet Instability."

12. J. Hansen and others, "Climate Change and Trace Gases," *Philosophical Transactions of the Royal Society* 365 (2007): 1925–54, doi: 10.1098/tsta.2007.2052. (http://pubs.giss.nasa.gov/docs/2007/2007_Hansen_etal_2.pdf).

13. James Hansen, "Scientific Reticence and Sea Level Rise," *Environmental Research Letters* 2 (March 23, 2007) (http://arxiv.org/ftp/physics/papers/0703/0703220.pdf).

14. Robert Zubrin, *Energy Victory* (Amherst, N.Y.: Prometheus Books, 2007), p. 109, quoting Louis D. Johnston and Samuel H. Williamson, "The Annual Real and Nominal GDP for the United States, 1790–Present," *Economic History Services* (July 27, 2007); see also http://www.measuringworth.com/datasets/usgdp/result.php.

15. Zubrin, *Energy Victory,* p. 109.

16. Ibid., pp. 109–10.

17. McKinsey Global Institute, "Curbing Global Energy Demand Growth: The Energy Productivity Opportunity," report (New York: McKinsey & Company, May 2007), p. 11.

18. IPCC, "Summary for Policymakers," in *Climate Change 2007: The Physical Science Basis.*

19. Pearce, *With Speed and Violence,* p. 59.

20. Romm, *Hell and High Water,* p. 86.

21. Ibid., quoting Peter Barrett.

22. Jay Gulledge, of the Pew Center on Global Climate Change, generated maps showing inundation patterns in May 2007. See appendix B in chapter 3 of this volume, pp. 89–91.

23. Roger Pielke and others, "Climate Change 2007: Lifting the Taboo on Adaptation," *Nature* 445 (2007): 597–98, citing Kelvin S. Rodolfo and Fernando P. Siringan, "Global Sea-level Rise Is Recognised, but Flooding from Anthropogenic Land Subsidence Is Ignored around Northern Manila Bay, Philippines," *Disasters* 30 (2006): 118–39.

24. J. L. González and T. E. Törnqvist, "Coastal Louisiana in Crisis: Subsidence or Sea Level Rise?" *Eos: Transactions of the American Geophysical Union* 87, no. 45 (November 7, 2006): 493–508.

25. Raymond S. Bradley and others, "Climate Change: Threats to Water Supplies in the Tropical Andes," *Science* 312 (2006): 1755–56.

26. Lawrence Wright, *The Looming Tower: Al Qaeda and the Road to 9/11* (New York: Knopf, 2006), p. 49.

27. National Research Council, *Making the Nation Safer: The Role of Science and Technology in Countering Terrorism* (Washington: National Academy of Sciences, 2002), p. 182.

28. John S. Foster Jr. and others, *Report of the Commission to Assess the Threat to the United States from Electromagnetic Pulse (EMP) Attack* (Washington: U.S. Electromagnetic Pulse Commission, 2004), pp. 1–2.

29. Ibid., pp. 17–18.

30. Ibid., pp. 40.

31. McKinsey Global Institute, "Curbing Global Energy Demand Growth," p. 12

32. The Rocky Mountain Institute (RMI) describes itself as "a nonprofit organization that fosters the efficient and restorative use of resources so that companies, governments and organizations are more efficient, make more money, and do less harm to the environment." For more information, see www.rmi.org.

33. Michael Kintner-Meyer and others, "Impacts Assessment of Plug-in Hybrid Vehicles on Electric Utilities and Regional Power Grids," Part I, "Technical Analysis," report prepared by the Pacific Northwest National Laboratory for the U.S. Department of Energy, November 2007 (www.pnl.gov/energy/eed/etd/pdfs/phev_feasibility_analysis_combined.pdf).

34. Willett Kempton and Jasna Tomic, "Vehicle-to-Grid Power Fundamentals: Calculating Capacity and Net Revenue," *Journal of Power Sources* (2005)(www.sciencedirect.com).

eight

Setting the Negotiating Table: The Race to Replace Kyoto by 2012

JULIANNE SMITH AND ALEXANDER T. J. LENNON

For the global response to climate change, 2007 was a landmark year. It began in January with President Bush's State of the Union address, in which he for the first time acknowledged "the serious challenge of global climate change," and concluded in December with the Bali Roadmap which global negotiators will use to seek to finalize an agenda for a framework by 2009 in Copenhagen to replace the Kyoto accord, due to expire in 2012. Although this was the ambitious officially declared agenda, Yvo de Boer, the executive secretary of the United Nations Framework Convention on Climate Change (UNFCCC), revealingly stated in an October 2007 interview, "I think the challenge in the next two years will be to design a climate policy that is good for the United Sates, good for China, and good for the EU."[1]

These three global powerhouses alone are responsible for roughly half of global greenhouse gas emissions (GHG), according to the World Resource Institute's Climate Analysis Indicators Tool (CAIT), emitting 20.4, 14.1, and 14.7 percent of global GHG emissions, respectively, in 2000, the most recent year for which all GHG emissions figures are available.[2] No other country is responsible for more than 5.7 percent. If these three players can agree, the

The authors both gratefully acknowledge Derek Mix, a fellow in the Center for Strategic and International Studies Europe program, for his substantial research and writing contributions to this chapter.

core of a global framework exists. The question is: Can they? In this chapter we examine the ways in which Europe, the United States, and China see the challenge of global climate change.

Europe's Leadership

Europe is rightly perceived as a global leader when it comes to climate change policy. The European Union was a central actor in the formulation and adoption of the UNFCCC, the first intergovernmental framework for addressing the issue, from 1992 to 1994. Over the next three years, the EU again played a crucial role in negotiating the Kyoto Protocol, which came into force in 2005. The Kyoto Protocol sets mandatory and legally binding targets for participating industrialized countries to reduce their greenhouse gas emissions by an overall total of 5 percent from 1990 levels by 2012. The protocol incorporates a number of flexible mechanisms to allow countries to meet their emissions reduction goals. These include national or regional emissions trading schemes and credits for sponsoring clean development projects or increasing carbon sinks, such as forests, either at home or in developing countries. By the time the Kyoto Protocol came into effect in early 2005, an internal EU Emissions Trading Scheme (ETS), the first international carbon-trading system, had already been set up. The ETS was established in October 2003 and came into operation in January 2005.[3]

The new Energy Policy for Europe (EPE), presented by the European Commission in January 2007 and approved by the spring 2007 European Council, makes it clear that responding to climate change is a top EU priority. The EPE commits the EU to reducing its greenhouse gas emissions by 20 percent by 2020 (compared to 1990) independent of what other countries decide to do, and pledges a 30 percent reduction should other developed countries commit to this level.[4] The EPE Action Plan calls for the EU, already the global pacesetter in renewable energy (with, for example, nearly two-thirds of the world's wind energy market), to triple its use of renewable energy sources by 2020 to provide for 20 percent of overall consumption. The plan additionally sets out, albeit in general terms, new regulatory measures to improve energy efficiency that highlight the importance of leveraging the internal European energy market and developing energy-saving and low-carbon technologies.[5]

The awareness and concern of European policymakers regarding climate change are reinforced by European public opinion: more than four-fifths of respondents to a Gallup Poll released in March 2007 agreed that they were "aware that the way they consume and produce energy in their country has a negative impact on climate," and 87 percent were either "very much concerned" or "to some degree concerned" about the effects of climate change and global warming.[6]

Within the overall goal of a 5 percent reduction in emissions by 2012, commitments under the Kyoto Protocol by individual countries vary. The fifteen EU member states who originally voted for the Kyoto Protocol in 1997 and eight of the ten central and eastern European states that joined the EU in 2004 and 2007 committed to the greatest reduction of all Kyoto Protocol participants: 8 percent for all the states combined. In deciding how the overall 8 percent reduction could be achieved, the EU-15 states assigned widely varying targets to each country. On one end, Germany and Denmark each committed to a 21 percent decrease in greenhouse gas emissions, while Greece and Portugal have ceilings under which emissions may increase no more than 25 percent and 27 percent, respectively.[7]

At the National Level

Although the European continent deserves kudos for its ability to match its rhetoric on climate change to tangible action, there are differences within Europe on how countries have chosen to address the challenge. For example, the size and composition of national industrial and transportation sectors make for differences in greenhouse gas emission levels and also in the type and level of adjustments a national economy can tolerate in the name of protecting the environment.

Similarly, individual countries have their own unique mixture of energy dependencies, in terms of both what their core sources are and where they come from. Thus, although an EU-wide consensus on the issue of climate change and the need to address it does indeed exist, there are also twenty-seven underlying national perspectives—not to mention those of non-EU European states such as Norway—on the importance of, and best solution to, the problem.

Germany is an important leader of the European charge on climate change policy and shoulders a substantial part of the burden. As Europe's largest economy, Germany's planned 21 percent reduction of carbon dioxide (CO_2) emissions by 2012 under Kyoto accounts for nearly three-quarters of the overall 8 percent EU reduction. With the ambitious commitments of the EPE, Europe is faced with the challenge of achieving a further 12 percent reduction between 2012 and 2020, and with its weighted portion factored in, Germany is looking at a total 40 percent reduction in CO_2 generation over a fifteen-year period.

Achieving such an ambitious goal requires a nearly holistic approach, linking a gradual overhaul of the way German industry operates and a society-wide commitment to changes in everyday lifestyle. This requires a strong emphasis on energy efficiency from the industrial level all the way down to household electrical appliances, greater use of renewable energy sources, and possibly such controversial measures as a blanket 130 kilometer-per-hour (80-mile-per-hour) speed limit on the autobahn, which in some stretches has no speed limit at all. Such changes have the potential to drastically reduce German CO_2 emissions even beyond Kyoto Protocol targets, as research on new technologies for environmentally friendly CO_2-free power plants and fuels continues.[8]

Supplementing its own national vision, Germany has put considerable effort into garnering more international support for climate change initiatives at a regional and global level. Germany's was the first-ever European government to include a Green Party in a governing coalition, and the Grand Coalition government of Chancellor Angela Merkel opted to push climate change and environmental issues as a key part of its agenda during its 2007 presidencies of the EU and the Group of Eight (G-8). The focus, as witnessed during the G-8 summit in Heiligendamm, has been on turning global concern into action. For Merkel, recognizing the severity of the problem is only the first step. The next step, which she continues to pursue with great commitment, is getting international actors, including the United States, China, and India, to agree to binding targets.

Under Prime Minister Tony Blair, the United Kingdom set about achieving its Kyoto commitment of a 12.5 percent emissions reduction by raising emissions standards for automakers, introducing a graduated auto tax based on fuel efficiency, and aiming to increase national use of biofuels. In March 2007 Blair also set a long-term national goal of a 60 percent CO_2 emissions reduction by 2050, which will be implemented through a series of five-year "carbon budgets."[9] Although it is debatable whether the United Kingdom is currently on track to meet the target for 2050, it is on track to fulfill its Kyoto commitment.

Yet the tactics of British climate change policy do split along party lines. The Labour Party stance emphasizes the importance of international agreements and the role of positive incentives to change behavior, such as lower taxes for environmentally friendly vehicles and buildings. The Conservatives, by contrast, advocate managing the issue through higher national taxation on emissions-causing behavior, such as emissions taxes on airline passengers and airplane fuel.

Although a consensus on the need to address climate change thus exists across the British political community, the governing party has a firm grasp on the reins of policy implementation in the British parliamentary system. Thus, Labour is free for the time being to work for emissions reduction within the strategy that the party has laid out. Meanwhile, the opposition Conservatives can be expected to continue to present alternative visions of how to combat climate change as they seek to differentiate their policies from those of Labour in the run-up to the general election that is expected to take place in the spring of 2009.

In general, the French government and public are in line with the European consensus regarding the importance of countering climate change. Initially, though, France did oppose the EPE because the country's nuclear power plants, which provide more than three-fourths of France's power needs, were excluded from national calculations of emissions responsibility. Once the EU agreed to take the French nuclear sector fully into account as a low-carbon energy source, France threw its wholehearted political backing behind the EPE.

Today, France can boast that its emissions have actually slightly decreased, even though France's assignment under Kyoto was simply to maintain emissions at 1990 levels. France is expected to play an even larger role in Europe's climate policy under President Nicolas Sarkozy, who has already made a number of pledges to strengthen his country's commitment to combating global warming. In his acceptance speech after his election win in May 2007, Sarkozy also urged the United States to show more leadership on tackling global warming.[10]

Shortcomings and Divisions in the EU

Despite Europe's laudable focus on climate change at the regional and national levels, fruitful action has not always followed the rhetoric. France, Sweden, and the United Kingdom are on track to meet or even exceed their Kyoto targets for CO_2 emissions reduction, but others, including Ireland, Portugal, and Spain, are badly behind.[11]

Although the ETS carries real symbolic importance, the first phase (2005 to 2007) has witnessed a number of serious shortcomings. At the start of the ETS, many targets for major emitters were set too high, and the allocation of carbon credits was far too generous. As a result, many large polluters were not required to reduce emissions, nor did they ever have to purchase credits, since many were already sitting on a surplus. When news of the credit hoards became public in the spring of 2006, the ETS market price for carbon credits collapsed.

Furthermore, under the Kyoto Clean Development Mechanism, European companies can trade credits outside of Europe, paying large sums to cash-hungry polluters in the developing world, especially China and India, for their carbon credits. But this influx of cash spurs expansion and new operations in the developing world, generating new emissions. Critics also argue that the money spent in the global emissions trading market—$30 billion in 2006—would have made a much greater difference had it instead been invested in emissions-reducing technologies.

In fact, the very existence of the market acts as a disincentive to many companies to change their polluting ways and move away from fossil fuels toward renewable energy sources and new technologies. With this first phase of the ETS admittedly a learning phase, the EU will need to apply its lessons vigorously to the second phase, 2008 to 2012, including setting stricter emissions limits and auctioning credits off, rather than handing them out.[12]

Although it is widely expected that Europe will continue to be a global leader in climate change policy, internal divisions on the continent do pose a number of potential problems. Intra-European east-west tensions flared during the European Council negotiations of the EPE. The economies of the new member states of central and eastern Europe are generally far more dependent on coal, gas, and CO_2-generating manufacturing than their counterparts in western Europe. Poland, for example, derives 90 percent of its energy from coal.[13]

These countries also have a much lower portion of renewable sources in their energy mix. Estonia's renewable energy sources account for 1 percent of energy sources, whereas Austria's account for 60 percent. These facts led the Czech Republic, Hungary, and Poland to oppose the EPE. They felt that the potential economic burdens of emissions reduction would be too great and the difficulty of meeting the renewable energy targets too extreme. A compromise was reached whereby the implementation of the EPE will mean more permissive emissions targets for the new members and possibly west-to-east subsidies of technology and energy supply.[14]

Intriguingly, diversification of energy supply through the development of renewable, alternative energy sources would be of greatest benefit to the central and eastern European EU members: many of these states currently are highly dependent on Russian oil and gas for their energy needs, and they are finding the affordability and availability of their energy sources increasingly vulnerable to the political aims of a Kremlin that boldly wields energy as an instrument of foreign policy. Thus, those countries with the least realistic capacity to diversify their energy and where economics still outweigh environmental concerns arguably have the greatest political rationale for seeking alternative sources.

Energy dependence is also an issue for western Europe, as the original EU-15 account for nearly 90 percent of the EU-25 gas market, which relies on Russia for 24 percent of its supply.[15] In addition, western Europe stands at the end of a pipeline network that runs out of Russia through eastern Europe, making it subject to disruptions anywhere along the way. These political considerations notwithstanding, the western European countries seem to be more inclined to diversify for environmental rather than political concerns. One can only hope for the gradual evolution of a state of affairs wherein all member states find it in their interest to pursue the same goals of emissions reduction and energy diversification seriously, even if their reasons differ widely.

The desirability and acceptability of nuclear power as a carbon-free energy source is another persistent topic of passionate debate in Europe. This issue has led to the creation of unlikely coalitions of interest, with pro–nuclear energy countries such as the Czech Republic, Finland, France, and Slovakia on one side and countries with broadly antinuclear publics, such as Austria, Denmark, and Ireland, on the other. Despite its appeal, some countries have already taken dramatic steps to reduce their reliance on nuclear energy. In a decision made under the Red-Green government of Chancellor Gerhard Schroeder, Germany plans to do away with its nuclear plants, which currently provide one-third of the country's power, by 2020. Ironically, this power supply will have to be replaced mainly by coal, which already accounts for more than one-half of Germany's electricity.

Finally, business leaders have predictably expressed concern that the EPE will hurt competitiveness and that it is unclear how the targets can be met. In January 2007 the heads of BMW, DaimlerChrysler, and Volkswagen sent a joint letter to the European Commission complaining that the EPE would unduly burden and harm the German auto industry. Although German carmakers have introduced some new technologies that reduce auto emissions and are gradually introducing hybrid vehicles, manufacturers often argue that significantly lower emissions limits simply cannot be met by most of the car models currently made by companies such as Audi, BMW, Mercedes, and Porsche.

Europe's Pivotal Role

The many shortcomings of the Kyoto Protocol are well known. It is in the nature of international agreements that there is no real enforcement mechanism to make the targets truly legally binding. Moreover, the Kyoto Protocol is limited to countries that are defined as industrialized, and key developing countries are not covered. This fact led the United States, the world's largest generator of CO_2, to refuse to ratify the agreement. Whereas the EU appears likely to meet its emissions reduction targets by 2012, growth in greenhouse gas emissions remains strong in Brazil, Canada, China, India, and the United States.

The future of managing climate change nevertheless rests with the next round of international agreement. With the Kyoto Protocol set to expire in 2012, the details of a regime to replace and build on it remain unclear. Many look to the EPE as setting the bar for a new international accord on climate change, which means European credibility is now on the line. Even if Europe achieves its internal goals to reduce greenhouse gas emissions, which it may very well do, it will only address a small portion of the problem.

With Europe's share of global pollution and energy consumption set to decline significantly over the next thirty years, the vision of the EU is now turning outward. According to projected scenarios decades in the future, the effects of climate change in Africa, the Middle East, South Asia, and elsewhere will radically impact Europe. As a result, European policymakers need to set a broad international negotiating strategy and get started on the far-ranging diplomacy needed to bring an aggressive post–Kyoto Protocol treaty into being. This means engaging the United States and developing countries such as China to bring these key players into the fold.

Changing U.S. Views

The global perception of the United States is that it is a laggard on climate change. Given its size and large contribution to global emissions, many countries around the world believe the United States should—and could—be doing more. Data from the Pew 2007 Global Attitudes project show that in thirty-four of the thirty-seven countries surveyed, the United States is named by a majority or a clear plurality as the country that is "hurting the world's environment the most."[16] That sentiment is shared by many Americans as well, with one-third of those surveyed rating their own country as the world's biggest polluter. For almost three decades, small groups of Americans have worked to promote climate policies, but to date the United States has shown very little leadership on this global challenge.

U.S. awareness of the potential problem of climate change first became widespread in the late 1980s. In 1986 and 1987, the climate expert James Hansen began expounding the view that global warming due to the greenhouse effect was to become a serious issue over the next 20 years.[17] Hansen's congressional testimony on the topic in the summer of 1988—a summer featuring severe droughts and heat waves—catalyzed the attention of the media, environmental groups, and the scientific community.

Hansen's efforts, however, were countered in 1989 when corporations from big industry, notably petroleum and automobile companies, founded the Global Climate Coalition. The sole purpose of this coalition was to refute any suggestion that action against the greenhouse effect was needed. Those views on the subject found a largely receptive audience within the George H. W. Bush administration. Yet, just before leaving office, Bush did sign on to the UN Framework Convention on Climate Change to counter his bad environmental reputation. Of course, the UNFCCC was not binding in any way, which made it easier for groups such as the Global Climate Coalition to accept.

Between the release of the first Intergovernmental Panel on Climate Change (IPCC) report in 1990 and the second report in 1995, international consensus on the severity of global warming gradually solidified and gained strength, particularly within the scientific community. Spearheaded by Vice President Al Gore, the Clinton administration pushed strongly for the Kyoto Protocol in 1997. Ironically, the United States fought to have a cap-and-trade system, something the United States first developed as part of the 1990 Clean Air Act, inserted into the protocol. In the face of mounting congressional opposition, however, the Clinton administration ultimately refused to sign the protocol.

The conservative-led Congress, supported and encouraged by groups such as the Global Climate Coalition, argued that adherence to the Kyoto Protocol would raise U.S. energy and gas prices and give other countries, such as India and China, an unfair economic advantage. After the U.S. Senate passed a declaration, 95–0, that Congress would not ratify the Kyoto Protocol unless it applied also to developing countries, the Clinton administration did not bother to submit the treaty for ratification.

The administration of George W. Bush has been pointedly skeptical on climate change and has introduced no legislation to deal with it.[18] In advance of the G-8 summit in June 2007, during some tense negotiations on whether or not the summit communiqués would include binding targets, Bush did invite the top fifteen emitters to attend a climate conference in Washington in late September 2007, following a UN meeting in New York on the same issue earlier that week.

The Washington meeting represented a significant breakthrough in the U.S. approach, as Bush acknowledged the importance of the issue, calling for the world's leading emissions producers to work together and set long-term emissions reduction goals in the context of a Kyoto Protocol successor treaty for 2012. Yet the presentations made by Bush and Secretary of State Condoleezza Rice also served to underline the wide gap between the thinking of policymakers in the United States and in much of the rest of the international community, with the United States continuing to oppose binding international treaties that contain minimum requirements and penalties for noncompliance. Rather, without suggesting concrete numbers, the administration proposed that each country should determine its own goals, to be pursued voluntarily, and that developing countries bear as much responsibility as developed ones. From the perspective of European policymakers, this episode represented on the one hand an important initial shift in a positive direction for the United States, yet on the other it demonstrated that U.S. policy on climate change remains largely isolated from and out of step with the worldview of Europe.[19]

Signs of Progress

Although the U.S. government has been dragging its feet on addressing climate change, there have been some shifts in U.S. policy in recent months. As one U.S. climate expert put it, "The United States is lacing up its running shoes and preparing to join the race."[20] Scientific evidence, support from businesses and industry, the promotion of climate-friendly policies as an element of faith, state and local initiatives, and the Democratic majority in Congress are enabling progress on this contentious issue.

First, the science has become both stronger and more visible. The IPCC's *Third Assessment Report*, published in 2001, provided the media, policymakers, the general public, and academics with much stronger evidence of a warming Earth (even though parts of the report were strongly contested). It also highlighted the role of greenhouse gas emissions. Perhaps most striking was the observable evidence, often via satellite imaging, that the report provided on the impacts of warming on the biosphere and on human societies. The *Fourth Assessment Report* in 2007 had an even greater impact, confirming with near certainty that carbon dioxide and other greenhouse gases from human activity are the main cause of global warming. Various extreme climate incidents, ranging from the European heat wave of 2003 to destructive storms such as Hurricane Katrina in 2005 to severe droughts and dwindling water resources in eastern Australia, have also provided skeptics in the United States and elsewhere with troubling firsthand accounts of the impact of warming on their societies.[21]

Second, increasing numbers of business leaders have gradually come to consider action on global warming as imperative in order to maintain energy security, economic growth and trade, and U.S. global leadership. Industry has also discovered that "going green," however vaguely defined, has considerable appeal among the public. Furthermore, businesses now see economic opportunities in new "green" technologies. Therefore, as the science of climate change advanced and grew in scope in the 1990s, and both the indirect and direct benefits of becoming environmentally friendly became more apparent, corporations began pulling out of the Global Climate Coalition, reducing the threat of the business veto on U.S. government action. In fact, many U.S. corporations are now serving as agents of change on this issue through efforts such as the U.S. Climate Action Partnership, which is a joint endeavor among large corporations such as Alcoa, BP America, Caterpillar, Duke Energy, DuPont, and GE and environmental groups to press for urgent action.

Third, many evangelical Christian groups have come to view combating climate change to be an obligation of faith. At first, these groups promoted individual responsibility to conserve. But some prominent church leaders have recently taken their cause to Washington, urging the federal government to take a more aggressive stance in confronting the issue. In early 2006, for example, a coalition of evangelical leaders issued "An Evangelical Call to Action," asking Congress and the Bush administration to restrict carbon-dioxide emissions.[22] That call triggered some fierce debates inside the evangelical community. But the increased attention on this issue among both evangelicals and a wide array of other religious groups, including Roman Catholics and Jews, has heightened awareness among the general public and caught the ears of Republican leaders in Congress and the administration.

Fourth, absent federal-level participation in Kyoto, the United States has witnessed a number of innovative approaches at the local and state levels. The best-known model is California, which has established a state Climate Action Team to devise greenhouse gas emissions reduction strategies based on both technology and regulation. Numerous businesses with operations in California, including DuPont and IBM, have voluntarily agreed to state emissions reduction targets. The state's motor vehicle plan aims to reduce car emissions, the greatest source of greenhouse gas emissions, by 30 percent by 2016. If the entire United States reduced its per capita emissions to California's level, U.S. pollution would be significantly lower than that called for by Kyoto.[23]

California is not the only state in the union showing muscle on this issue. Twelve other states have adopted caps on auto emissions, and 435 U.S. mayors, both Republicans and Democrats, have signed the U.S. Mayors Climate Protection Agreement, whereby they commit their cities to meeting the Kyoto emissions targets.[24] In another sign that climate change is no longer associated with those on the left, Jon Huntsman Jr., the Republican governor of conservative Utah, has become a cap-and-trade advocate and committed himself to working with California on reducing carbon emissions. Dan Schnur, a Republican political analyst, called that shift "the energy equivalent of Nixon going to China."[25]

Finally, the Democratic takeover of Congress in 2006 has also advanced climate change debates in Washington. According to a recent Zogby International postelection survey, one-half of Americans who voted in the 2006 midterm elections said concern about global warming made a difference in their vote.[26] A handful of global warming skeptics lost influential posts in that political transition, including the chairman of the Senate Environment and Public Works Committee, Senator James Inhofe (R–Okla.); Inhofe has called global warming "the greatest hoax ever perpetrated on the American people."[27] He was replaced as committee chair by Senator Barbara Boxer (D-Calif.), an outspoken critic of the administration, particularly on climate-related issues.

Democrats achieved one of their major climate and energy goals with the passage of the Energy Independence and Security Act in December 2007. Among its major provisions, the act sets higher fuel economy standards for vehicles in the United States for the first time in twenty-two years, requires a fivefold increase in production of renewable fuels by 2022, and establishes new efficiency requirements for household appliances and government buildings, seeking to phase out the incandescent light bulb within the next decade. Nevertheless, in order to pass in the Senate and avoid a presidential veto, the bill was watered down, losing its initial $13 billion tax increase on oil companies—designed to subsidize solar, wind, and geothermal energy projects—as well as requirements for greater use of renewable resources in national power generation.

Many Democrats and some Republicans, including Senators John Warner of Virginia and John McCain of Arizona, now support a cap-and-trade system that would allow industries that fall under a mandated emissions cap to trade credits to those that do not.[28] Some hope that as Bush looks to build a legacy that reaches beyond Iraq and the war on terrorism, he will become increasingly accommodating on adopting mandatory controls. Environmentalists are also hopeful that the 2008 presidential election will bring increased attention to climate issues, and that the presidential candidates will make climate a more central part of their political platforms and of their presidency.

Evolving U.S. Public Opinion

These scientific, business, political, and religious shifts have been accompanied by shifts in public opinion. According to a survey conducted in February 2007, the percentage of Americans who say global warming is a serious problem has risen to 83 percent, from 70 percent in 2004.[29] Some argue that the success of Al Gore's film, *An Inconvenient Truth,* which won the Oscar award for best documentary in 2007, has also heightened awareness of the dangers of climate change.

A number of skeptics, however, continue to question the science and oppose policy changes regarding climate change. Some claim that climate change is not taking place at all and that warming is simply a natural cycle of change that is not due to human activity. If Hurricane Katrina forced some skeptics to rethink their assumptions about the severity of the threat, the unusually cold winter in 2006–07 was cited as further evidence of the uncertainty of the problem. A sizable portion of the U.S. population continues to believe that changing human behavior will have no effect on the process whatsoever. Instead, humans must simply adapt to changing circumstances.

Caught between climate change advocates and the skeptics are those who admit that warming is occurring but oppose any initiative that might hurt the U.S. economy. These individuals, recognizing that the United States is the world's largest per-capita source of greenhouse gases, argue that the United States will pay the highest price for change. For example, if the United States were to put in place a cap-and-trade system, operating costs for U.S. firms would rise, making imported goods, especially from India and China, even more competitive and possibly driving U.S. companies out of business. Any solution must therefore include these rapidly developing countries.

Bridging the Gap

The challenge for the United States in the coming years will be to find a way to bridge the gaps between those who support mandatory cap-and-trade programs and those who advocate adopting alternative solutions, such as voluntary targets, and persuading as many skeptics as possible to alter their views on the science. This will be difficult for many reasons. First, time is short. With the Kyoto Protocol expiring in 2012, the United States and the broader international community do not have much time to begin the arduous task of reaching a global consensus on a post–Kyoto Protocol agreement. Many believe that, at the very latest, negotiations would need to start in 2009, when a new U.S. president comes into office. Yet climate experts also often concede that the United States is unlikely to sign an international treaty before domestic legislation is in place, which will certainly take more time than the pending Kyoto Protocol deadline provides.

Second, although U.S. public concern about this issue appears to be growing, Americans, even those who support the science, simply do not feel the same sense of urgency as others around the world. A common but false assumption is that the impact of climate change will spare the industrialized world, especially the United States, Europe, and Australia.[30] This makes building and maintaining the required momentum on this issue perhaps more challenging on U.S. soil than in other corners of the world.

In reality, a U.S. commitment to future climate change regimes will be essential to the regimes' success. As the world's greatest producer of greenhouse gases and the world's largest consumer of energy, any solution to this challenge must include the United States. Without it, any hope of bringing developing countries such as China on board is futile.

China's Awakening

While the United States has been viewed as a laggard in recognizing its responsibilities vis-a-vis climate change, China has been portrayed as distracted or even as avoiding the problem entirely. Obsessed with economic development, and unwilling to assume responsibility for what the Chinese view as a situation created by over a century of developed-world industrialization, Beijing has been perceived as a country that will not be bothered with climate change negotiations, at least until the developed countries reach their own serious agreements. Yet, like U.S. views, Chinese views of the climate change challenge and what to do about it have been changing.

The View from Beijing

Despite China's need to "catch up" economically with the developed world, Chinese perceptions have evolved on the risks involved and what to do about it. Beijing has begun to see climate change itself as a potential drain on the Chinese economy and a source of popular instability. Yet as a developing country with over three times as many people as Europe and four times as many as the United States, China views at least the metrics of climate change differently from those countries. China's greenhouse gas emissions have dramatically risen with its astounding economic growth since 1979, but Beijing takes issue with three ways in which emissions are typically measured.

First, annual current emissions or projections dominate most discussions, but instead, Beijing cites cumulative historical emissions to assess who is responsible for the problem. According to the World Resource Institute's Climate Analysis Indicator Tool (CAIT), the United States and EU emitted over 55 percent of the carbon dioxide from 1850 to 2003, whereas China was responsible for less than 8 percent. From China's perspective, those who are most responsible for causing the problem in the *past* should make the greatest sacrifices to address it.

Second, while others look at total national greenhouse gas emissions, Beijing cites its low *per capita* GHG emissions. On this measure China ranked hundredth in the world in 2000, according to the CAIT. From its perspective, China should not be penalized simply for having a lot of people, particularly when it is already faced with the challenge of economically providing for them.

Finally, Beijing also focuses on efficiency, or energy intensity, the ratio of energy consumption to GDP, and consequently to emissions intensity, or the ratio of carbon dioxide–equivalent emissions to GDP. From 1980 to 2000, while the value of China's economy quadrupled, its energy consumption only doubled, so it improved its energy intensity dramatically.[31] Collectively, these measures lead Beijing to the position, as expressed by the Chinese Foreign Ministry spokeswoman Jiang Yu, that "the key issue of the current international negotiations on climate change is that developed countries must continue to take the lead in cutting emission of greenhouse gases."[32]

While China's views on climate are moderating, its top priority, quite simply, remains economic growth, which officials cite as a "right" for developing countries. At the September 24, 2007, UN High-Level Event on Climate Change, Foreign Minister Yang Jiechi introduced his remarks by saying, "Climate change is an important development issue."[33] In 2006 Chinese state councilor Hua Jianmin also at least sought to deflect international pressure, making the converse argument that "economic development is not only a prerequisite for the subsistence and progress of human beings, but also a material foundation for the protection and improvement of the global environment."[34]

What has changed recently, however, is that China has begun to see the consequences of environmental damage generally, and climate change specifically—drought, crop shortages, and typhoons—as threats to economic growth. A February 2007 Lehman Brothers report cited estimates by Chinese researchers that environmental pollution in 2004 cost the Chinese economy 3.1 percent of GDP.[35] At the end of 2006, the Chinese Ministry of Science and Technology claimed specifically that "global climate change has an impact on the nation's ability to develop further."[36] That change in perspective means that China no longer views climate change and economic growth as mutually exclusive priorities, but as somewhat intertwined.

Chinese concerns about the environment transcend economic concerns and extend to the maintenance of social stability itself, instability being a terrifying prospect for the Chinese leadership. Across the country thousands of Chinese have already rioted, as in the 2005 incidents in Huaxi village and in Xinchang (see also chapter 4 of this volume, by Jonathan Podesta and Peter Ogden). Even though both of these demonstrations targeted just factory pollution, they showed that the Chinese people are willing to speak out and to take action over access to clean water.

Chinese officials are aware of the threat that global warming represents for their country. In early 2007 the deputy director of China's office of Global Environmental Affairs in the Ministry of Science and Technology, Lu Xuedu, pointed out that climate change will cause river levels to decline and droughts and floods to increase, and he warned specifically that demand would outstrip the supply of water in western China by up to 20 billion cubic meters (700 billion cubic feet) between 2010 and 2030. Qin Dahe, an expert at the Chinese Academy of Sciences, also raised the concern that glaciers on the Qinghai-Tibet Plateau could shrink from 500,000 square kilometers (195,000 square miles) in 1993 to 100,000 square kilometers (39,000 square miles) in 2030, reducing the meltwater that feeds many major rivers in Asia and jeopardizing the water supply for up to a billion people.[37]

A columnist for the London *Financial Times,* Gideon Rachman, put it best: "The government in Beijing faces a dilemma. Terrified of social unrest, it is reluctant to do anything that might slow economic growth—such as stopping the building of coal-fired power stations. Yet, water shortages are already causing social unrest in the country side and the water table is falling fast in Beijing."[38] Clearly, what China is willing to do to mitigate or adapt to climate change will be shaped by economic pressures, but climate change is now viewed not just as an environmental issue but also as an economic problem and a threat to political and social stability.

Coming to Terms with Climate Change

To combat climate change, China's National Development and Reform Commission released its first National Climate Change Program in June 2007, on the eve of the G-8 summit. The sixty-page document outlines China's guidelines, six basic principles, specific objectives, and policies to mitigate and adapt to the threat. The programs mentioned are mostly an amalgamation of policies implemented throughout the economy, particularly in the energy sector, that have the effect of reducing GHG emissions, even if they have been instituted for other reasons. For example, the plan claims credit for the potential of its one-child policy to reduce the number of the Chinese people who will contribute to carbon emissions.

One of China's six key principles is "to place equal emphasis on both mitigation and adaptation." Beijing has expressed that, as a developing country, it is most immediately concerned about China's resilience and ability to adapt to and manage climate-induced strains. While mitigation is an important element, it "is a long and arduous challenge, while adaptation to climate change is a more present and imminent task."[39] Adaptation techniques include adjusting cropping systems, developing stress-resistant crop varieties and biotechnologies, preventing desertification by building artificial grasslands, enhancing water resource management, promoting antiflood engineering systems, and constructing and expanding mangroves and other coastal trees and shrubs to protect against sea level rise.[40]

To mitigate climate change, China's focus remains on improving energy efficiency, increasing the use of renewable energy and alternate energy sources—such as hydropower, thermal power generation, and nuclear power—and reviving the gains it had made in energy intensity through 2002. Beijing has set ambitious targets, such as reducing emissions by 20 percent by 2010, but it faces a severe challenge in getting provincial and local officials to enforce these measures. To date, Beijing has not issued genuine incentives to provincial leaders and businesses to cut emissions, or penalties for failure to do so, or to adopt environment-friendly technologies. Recently, however, Beijing has begun to evaluate local leaders on energy-efficiency improvements and has had some initial success.[41] It may even consider how to use nongovernmental environmental groups as watchdogs on the activities of provincial and local leaders without threatening political stability.

Internationally, China will continue to feel the pressure to deal with the effects of climate change, as occurred when it was unable to significantly respond to regional calls for help after the Indian Ocean tsunami, and to be a "responsible stakeholder" in global negotiations to construct a post-Kyoto framework. China's international position is embodied in the Asia-Pacific Partnership for Clean Development and Climate (AP6 for short), a six-country initiative agreed to in July 2005 by Australia, China, India, Japan, South Korea, and the United States. The AP6 encourages new technologies that address climate change while promoting "sustainable economic growth and poverty reduction." Although the countries are criticized for not actually funding the agreement, its principle is still clear: do not regulate current businesses, but create new ones through incentives to develop and transfer cleaner, more efficient technology.

The June 2007 National Climate Change Program emphasizes that one of its six guiding principles to respond to climate change is "to rely on the advancement and innovation of science and technology. . . . China will make great efforts to develop new and renewable energy technologies and new technologies of energy conservation, to promote carbon sink and other adaptive technologies, to accelerate scientific and technological innovation and importation, and to provide a strong scientific support to address climate change and promote the capacity of sustainable development."[42]

In the view of China's leaders, the transfer of such technology is a critical component of any post-Kyoto framework. In a global opinion poll on views of global warming, China's population was among the most concerned: 83 percent responded that global climate change should be addressed, of whom 42 percent felt that global warming is a "serious and pressing problem" and that immediate action should be taken, and 41 percent thought that it "should be addressed, but its effects will be gradual, so we can deal with the problem gradually by taking steps that are low in cost." When asked what they would do if developed countries provided substantial aid, 79 percent said they would agree to limit emissions.[43] The National Climate Change Program reinforces the view that China needs technological assistance for a multitude of capabilities in order to effectively monitor, mitigate, and adapt to climate change.[44]

What Now?

Any viable solution to the challenge of climate change rests on the ability of the international community—particularly Europe, the United States, and

China—to engage with one another to combine strengths, experiences, and perspectives into a post-Kyoto framework. The year 2007 marks the beginning of that engagement. The global negotiations can be envisioned as a line of standing dominoes, each either pressuring or hiding behind the other. At the front of the line stands Europe, leaning on the United States, which to date has resisted approaches such as mandatory caps. Behind the United States stands China. Realistically, with U.S. presidential elections to be held in November 2008, negotiators can make progress at the working level, but political leaders will be anxiously awaiting what happens next. If a new U.S. administration takes a different stance toward, for example, some form of a carbon cap-and-trade system, and the proverbial U.S. domino falls, it will lean on China and pressure will pass to Beijing. Behind China stand other developing countries such as India and Brazil, the countries of the Middle East, and other developed countries such as Russia, Japan, and Canada, all of whom are undergoing their own national evaluations but will be influenced by the global line of dominoes.

One of the new ways in which individual dominoes are influenced by the global climate debate is that communities with little traditional connection to environmental and climate issues are being integrated into the debate. So far, a handful of studies, including this one, have been undertaken to bridge the gap between the national security and climate communities so that global warming receives the attention it warrants. Climate change will have major ramifications for migration, force posturing, failed states, and federal resourcing. The sooner national governments recognize climate change as a genuine and pressing national security issue, the faster it will receive the intellectual, financial, and diplomatic resources it merits.

Notes

1. "Interview: UN Official Calls on U.S., Australia to Join Climate Change Talks," Xinhua News Agency, October 4, 2007.

2. World Resources Institute, Climate Analysis Indicators Tool (http://cait.wri.org).

3. Peter Goldmark and Ernst von Weizsäcker, "The Decarbonization Challenge: U.S. and European Perspectives on Climate Change," part 2, Bertelsmann Stiftung Transatlantic Thinkers Series (Brussels: March 7, 2007), p. 6 (www.euractiv.com/29/images/DecarbonizationChallenge_tcm29-162797.pdf.

4. Council of the European Union, "Presidency Conclusions—Brussels, 8/9 March 2007," 7224/1/07/REV1 (Brussels: May 2, 2007), p. 12 (available at www.consilium.europa.eu).

5. European Commission, "Energy for a Changing World," March 2007 (http://ec.europa.eu/dgs/energy_transport/index_en.html).

6. Gallup Organization, "Attitudes on Issues Related to EU Energy Policy," *Flash Eurobarometer,* no. 206a (March 2007) (http://ec.europa.eu/public_opinion/flash/fl206a_en.pdf).

7. UN Framework Convention on Climate Change, "Countries Included in Annex B to the Kyoto Protocol and Their Emissions Targets" (http://unfccc.int/kyoto_protocol/background/items/3145.php) and "A Summary of the Kyoto Protocol" (http://unfccc.int/kyoto_protocol/background/items/2879.php).

8. "Europa muss eine Vorreiterrolle spielen" [Europe must play a pioneering role], interview with Chancellor Angela Merkel, *Sueddeutsche Zeitung,* March 5, 2007 (www.sueddeutsche.de/deutschland/artikel/331/104227); "Die Politik drückt sich um die Auseindersetzung mit der Industrie" [Politics is avoiding a debate with industry], interview with German environment minister Sigmar Gabriel, *Sueddeutsche Zeitung,* March 3, 2007 (www.sueddeutsche.de/wissen/artikel/246/104142).

9. George Monbiot, "Just a Lot of Hot Air," *Guardian,* March 5, 2007 (www.guardian.co.uk/g2/story/0,,2026711,00.html).

10. Nicolas Sarkozy, "Acceptance Speech," About.com (http://worldnews.about.com/od/presidentialelection/a/sarkozy_speech.htm).

11. "From Free Trade to Deep Integration," *EU Monitor 45,* April 18, 2007, p. 31.

12. Goldmark and Weizäcker, "Decarbonization Challenge"; Emily Flynn Vencat, "The Carbon Folly," *Newsweek,* March 12, 2007 (www.msnbc.msn.com/id/17435875/site/newsweek).

13. Dan Bilefsky, "EU Drafts Compromise to Fight Climate Change," *International Herald Tribune,* March 10, 2007 (www.iht.com/articles/2007/03/09/news/eu.php).

14. Ibid.; "Green Grind," TimesOnline, March 10, 2007 (www.timesonline.co.uk/tol/comment/leading_article/article1494964.ece); Bilefsky, "EU Drafts Compromise to Fight Climate Change."

15. "Statistics 2005," *Eurogas* (www.eurogas.org/uploaded/statistics%202005.pdf).

16. Pew Research Center, "34 Nations Call U.S. Biggest Threat to Environment" (http://pewresearch.org/databank/dailynumber/?NumberID=343).

17. Spencer Weart, "The Public and Climate Change," July 2007 (www.aip.org/history/climate/public2.htm).

18. Spencer Weart, "The Discovery of Global Warming," July 2007 (www.aip.org/history/climate/index.html).

19. "Warming to the Environment?" *Economist,* September 28, 2007, online edition (www.economist.com); Fiona Harvey and Andrew Ward, "Bush Calls for Fresh Global Approach to Climate Change," *Financial Times,* September 29, 2007, p. 6.

20. Peter Goldmark, "Time to Act Together: How Europe and the United States Can Collaborate on Climate Change and Energy Independence," speech, German Marshall Fund, Washington, D.C. (May 24, 2007).

21. IPCC, *Climate Change 2001: Synthesis Report: Third Assessment Report of the Intergovernmental Panel on Climate Change,* edited by Robert T. Watson (Cambridge University Press, 2001); IPCC, *Climate Change 2007: Impacts, Adaptation and Vulnerability: Contribution of Working Group II to the Fourth Assessment Report of the Intergovernmental Panel on Climate Change,* edited by M. L. Parry and others (Cambridge University Press, 2007).

22. Karen Breslau and Martha Brant, "God's Green Soldiers," *Newsweek,* February 13, 2006, p. 49 (www.msnbc.msn.com/id/11179145/site/newsweek).

23. State of California, Climate Action Team, "Fact Sheet: California's Program to Reduce the Impact of Global Warming," California Climate Change Portal (www.climatechange.ca.gov/climate_action_team/factsheets/2005-06_CAT_Q+A.PDF).

24. Anne Underwood, "Mayors Take the Lead," *Newsweek,* April 16, 2007, online edition (www.msnbc.msn.com/id/17996836/site/newsweek).

25. Michael Gardner, "Governor Acquires Unlikely Ally in Regional Global Warming Fight," Copley News Service, May 20, 2007 (www.signonsandiego.com/news/politics/20070520-9999-1n20utah.html).

26. See "Zogby Post-Election Poll: Dems Gained from Global Warming Debate" (www.zogby.com/news/ReadNews.dbm?ID=1194).

27. See "Climate Change Update: Senate Floor Statement by U.S. Sen. James M. Inhofe (R-Okla.), January 4, 2005" (http://inhofe.senate.gov/pressreleases/climateupdate.htm).

28. Frank Davies, "Congress Stymied on Global Warming Bills," *San Jose Mercury News,* July 23, 2007 (www.mercurynews.com/lifestyle/ci_6441203).

29. Global Strategy Group, "2007 Environment Survey: Key Findings," March 5, 2007 (www.loe.org/images/070316/yalepole.doc).

30. See Jay Gulledge, "Three Plausible Scenarios of Future Climate Change for Security Risk Assessment," working paper (Washington: Center for Strategic and International Studies, CSIS Climate Security Project, July 12, 2007), pp. 4–5.

31. Joanna I. Lewis, "China's Strategic Priorities in International Climate Change Negotiations," *Washington Quarterly* 31, no. 1 (Winter 2007–08): 155–74.

32. Jiang Yu, "China Says Developed Countries Must Take Lead in Cutting Greenhouse Gases Emission," *Xinhua,* February 6, 2007.

33. "Statement by H. E. Mr. Yang Jiechi at the Thematic Plenary on Adaptation of the United Nations High Level Event on Climate Change, New York, September 24, 2007" (www.chinaconsulatesf.org/end/xw/t366299.htm).

34. Hua Jianmin, "New Climate Partnership Needs Concrete Measures: Chinese State Councilor," *Xinhua,* January 12, 2006.

35. John Llewellyn, "The Business of Climate Change: Challenges and Opportunities," report (London: Lehman Brothers, February 2007).

36. Peter Harmsen, "Dire Warnings from First Chinese Climate Change Report," Agence France-Presse, December 27, 2006.

37. Ding Yimin, "China Goes on Offensive to Combat Climate Change," *Xinhua,* October 4, 2007.

38. Gideon Rachman, "Climate Change Is Not a Global Crisis—That Is the Problem," *Financial Times,* April 16, 2007.

39. People's Republic of China, National Development and Reform Commission, *China's National Climate Change Programme* (Beijing: June 2007), p. 24.

40. Ibid., pp. 27–28, 47–52.

41. Lewis, "China's Strategic Priorities in International Climate Change Negotiations," p. 160. Also see Jared Diamond, *Collapse: How Societies Choose to Fail or Succeed* (New York: Penguin Books, 2005), p. 375.

42. People's Republic of China, National Development and Reform Commission, *China's National Climate Change Programme,* p. 25.

43. See World Public Opinion.org, "Poll Finds Worldwide Agreement That Climate Change Is a Threat," March 13, 2007 (www.worldpublicopinion.org/incl).

44. People's Republic of China, National Development and Reform Commission, *China's National Climate Change Programme,* pp. 60–61.

nine
Conclusion: The Clear Implications of Global Climate Change

KURT M. CAMPBELL AND RICHARD WEITZ

This volume was the product of a year of collaboration and discussion among a new community of scientists, climate experts, and foreign policy and national security practitioners. As our work came to a close, Al Gore and the IPCC were awarded the 2007 Nobel Peace Prize for their efforts to raise public awareness of climate change and its daunting implications for global security. Much has been made of the fact that the award was for peace rather than science or in some other field, but in awarding Gore and the IPCC the Nobel Peace Prize, the judges showed that they understood precisely what the stakes are when it comes to global climate change. Nothing could better underscore that the time is right: there is an urgent need to understand the nature of the climate change challenge and, more to the point, there is an urgent need for all Americans to appreciate the need for action to confront this global challenge.

Our diverse group undertook a scenario exercise in hopes of reaching a better understanding of the consequences the world could realistically face from climate change across the range of plausible effects. Our intention was to influence the public debate about climate policy. We came away with considerable clarity in our own minds: the United States can expect that climate change will exacerbate already existing North-South tensions, dramatically increase global migration both inside and between nations (including into the United States), lead to increasingly serious public health problems, heighten interstate tension and possibly conflict over resources, collapse agricultural markets and global fisheries, challenge the institutions of global governance, cause potentially destabilizing domestic political and social repercussions, and spur unpredictable shifts in the global balance of power, particularly where China is concerned. The state of humanity could be altered in ways that create strong moral dilemmas for those charged with wielding national power, and also in ways that may either erode or enhance America's place in the world.

Taken together or even one at a time, some of these challenges have the potential to overwhelm national governments and international institutions. It is difficult to anticipate just how these cascading calamities might ultimately unfold, but the prospects for destabilizing global effects are clearly on the horizon. The overwhelming message is that early steps to limit or mitigate climate change are essential because later, longer-term efforts to adapt or anticipate may not be possible.

As table 9-1 summarizes and this report makes clear, climate change has the potential to be one of the greatest national security challenges that this or any other generation of policymakers is likely to confront.

Although our charge was to offer projections based on scientific modeling, rather than predictions, the *expected* climate change scenario in this report is appropriately named and can be taken as a minimum basis for planning. As John Podesta and Peter Ogden write in chapter 4, the environmental effects in this scenario are "the least we ought to prepare for." National security implications include heightened internal and cross-border tensions caused by large-scale migrations; conflict sparked by resource scarcity, particularly in the weak and failing states of Africa; increased disease proliferation, which will have economic consequences; and some geopolitical reordering as nations adjust to shifts in resources and prevalence of disease. Oil and natural gas exporters might gain an upper hand, while energy importers will suffer geopolitically. All these things could cause the internal politics of nations to radicalize or destabilize. Across the board, the ways in which societies react to climate change will refract through underlying social, political, and economic factors.

In the case of *severe* climate change, projected massive nonlinear events in the global environment give rise to massive nonlinear societal events. In this scenario, nations around the world will be overwhelmed by the scale of change and by pernicious challenges, such as pandemic disease and water and food shortages. The internal cohesion of nations will be under great stress, including in the United States, owing to a dramatic rise in migration, changes in agricultural patterns and water availability, and the pulling away of wealthier members of society from the rest of the population. Protests,

Table 9-1. Summary of Key Environmental and National Security Implications of Three Climate Scenarios

	Scenario 1: Expected Climate Change	Scenario 2: Severe Climate Change	Scenario 3: Catastrophic Climate Change
Summary of climate change assumptions	Average 1.3°C warming .23 meter of sea level rise Approximately 30-year time frame	Average 2.6°C warming .52 meter of sea level rise Approximately 30-year time frame	Average 5.6°C warming 2.0 meters of sea level rise Approximately 100-year time frame
Key selected environmental stresses, based on scenario assumptions	Water scarcity affects up to 1.7 billion people Changed distribution of some infectious disease vectors & allergenic pollen species Up to 3 million additional people at risk of flooding Up to 30 million more people at risk of hunger due to crop failure	Water scarcity affects up to 2 billion people Increased burden from malnutrition, diarrheal, cardio-respiratory, and infectious diseases Up to 15 million additional people at risk of flooding Changes in marine and ecosystems due to weakening of the meridional overturning circulation	Water scarcity affects 3.2 billion people Increased morbidity and mortality from heat waves, floods, and droughts Approximately 30 percent loss of coastal wetlands Up to 120 million more people at risk of hunger due to crop failure Possible collapse of the meridional overturning circulation
Possible national security implications, based on scenario assumptions	Conflict over resources due to and driving human migration Immigrants—or even simply visitors—from a country in which there has been a significant disease outbreak may not be welcomed and could be subject to quarantine and lead to loss of national income. Dissatisfaction with state governments could radicalize internal politics and create new safe havens in weak and failing states	Wealthiest members of society pull away from the rest of the population, undermining morale and viability of democratic governance Global fish stocks may crash, enmeshing some nations in a struggle over dwindling supplies Governments, lacking necessary resources, may privatize water supply; past experience with this in poor societies suggests likelihood of violent protest and political upheaval	Migration toward U.S. borders by millions of hungry and thirsty southern neighbors is likely to dominate U.S. security and humanitarian concerns A shrinking Russian population might have substantial difficulty preventing China from asserting control over much of Siberia and the Russian Far East; the probability of conflict between two destabilized nuclear powers would seem high

(continued)

Table 9-1. Summary of Key Environmental and National Security Implications of Three Climate Scenarios (*continued*)

	Scenario 1: Expected Climate Change	Scenario 2: Severe Climate Change	Scenario 3: Catastrophic Climate Change
	A strengthened geopolitical hand for natural gas–exporting countries and, potentially, biofuel-exporting countries; a weakened hand, both strategically and economically, for importers of all fuel types Social services will become increased burden on central government where available The regional positions of Turkey and others will likely be strengthened as a result of the water crisis	Globalization may end and rapid economic decline may begin, owing to the collapse of financial and production systems that depend on integrated worldwide systems Corporations may become increasingly powerful relative to governments as the rich look to private services, engendering a new form of globalization in which transnational business becomes more powerful than states Alliance systems and multilateral institutions may collapse—among them the UN, as the Security Council fractures beyond compromise or repair	Rage at government's inability to deal with the abrupt and unpredictable crises Religious fervor, perhaps even a dramatic rise in millennial end-of-days cults Hostility and violence toward migrants and minority groups Altruism and generosity would likely be blunted U.S. military's worldwide reach could be reduced substantially by logistics and the demand of missions near our shores Electricity generation and distribution highly vulnerable to attack by terrorists and rogue states

civil unrest, and violent upheaval of governments are possible. The flooding of coastal communities around the world, especially in the Netherlands, the United States, South Asia, and China, has the potential to challenge regional and even national identities. Armed conflict between nations over resources and even territory, such as the Nile and its tributaries, is likely, and even a rise in nuclear tensions is possible. Globalization of the economy could halt and alliances could collapse.

The *catastrophic* scenario, in which average global temperatures increase by 5.6°C (10.1°F) by 2100, is by far the most difficult future to visualize. The author of this scenario (see chapter 6) notes that intense hurricanes will become increasingly common, as will droughts, floods, wildfires, heat waves, and churning seas. Hundreds of millions of thirsty and starving people will have to flee or perish, leaving the globe dotted with ghost towns. The abrupt and sudden nature of many of these phenomena will challenge the ability of all societies to adapt, including that of the United States. Two countervailing trends might provide a glimmer of hope in the long term: the most profound will be innovations in science and technology, and the social changes that come with them. New technologies may dramatically affect global energy consumption and give us more tools to mitigate the effects that can no longer be prevented. Second, the human capacity for hope and determination should never be discounted, even in the most extreme of circumstances. As the effects of climate change become undeniable, the will of people to find collaborative solutions may provide the much-needed catalyst for global change.

These scenarios aim not to speculate on events centuries in the future, as some scientific models do, but to consider plausible developments using a reasonable time frame for making acquisition decisions—such as a fighter plane or a major war ship—or judgments about larger geopolitical trends. In national security planning, it generally can take about thirty years to design a weapons system and bring it to the battlefield, so it is important to anticipate future threat environments and to begin preparations now.

Implications of Climate Change

Although the scenarios are certainly interesting and thought provoking in and of themselves, the point of such an exercise is to draw observations and lessons that can help guide decision making in the present tense. In that sense, there are ten highly consequential implications of climate change that can be drawn from this report that, it is hoped, will animate dialogue and debate now, when consequential change is still possible, rather than waiting for an uncertain future to take actions.

1. Increase in North-South Tensions

North-south tensions are likely to increase as problems of equity arise in the entire climate debate. Developing countries will generally suffer most from adverse impacts of global climate change because of their limited response capacities. Their ability to triumph through several such events simultaneously or in succession is even less likely. In contrast, wealthier countries (and their most wealthy citizens) appear better positioned to cope with at least early consequences of modest climate change. To add to this tension, the wealthiest nations became so in no small part by burning the very fossil fuels responsible for increasing the concentration of atmospheric greenhouse gasses. A failure of the developed nations to assist developing countries to manage the climate change challenge will almost certainly cause a further spike in north-south tensions.

2. Increased Migration

A profound increase in the movement of climate refugees will cause greater tensions and perhaps violent conflicts between and within countries over uncontrolled immigration issues. Such massive migrations within a relatively short time are likely to be deeply problematic for the "host" countries for these climate refugees. In the Western Hemisphere, Americans may find themselves struggling to resettle tens of millions of their own citizens, driven by high water from the Gulf of Mexico, South Florida, and much of the East Coast reaching nearly to New England. Under severe scenarios, climate-induced migration could transform the ethnic character of major countries and world regions, especially the European Union. An influx of Muslims into Europe, for example, could lead to new tensions over foreign policy priorities toward Muslim countries or Islamist terrorism. Historical reactions to natural disasters, such as public rage at government's inability to deal with the abrupt and unpredictable crises, increased religious fervor, and hostility and violence toward migrants and minority groups, could dramatically worsen perennial tensions about immigration.

3. Health Consequences

Climate change will have profoundly negative consequences for global health, especially in poorer regions of the world. Not only will some areas become more hospitable for vector-borne diseases, but any climate-induced shortages in local food and water supplies will increase the population's susceptibility to illness. The resulting increase in deadly or debilitating diseases could worsen poor economic conditions in the affected regions by decreasing worker productivity, requiring governments to spend more on public health rather than other priorities, and limiting tourism in those areas that have a tourist industry.

4. Water Scarcity

Over the next three decades, climate change–exacerbated water scarcity could well contribute to instability in many regions of the world—a dire problem in itself that may also be a detriment to agriculture and basic subsistence. Tensions could increase within and between states that experience shrinking water supplies; countries with an abundant water supply could seek to exploit it for economic, political, and diplomatic advantage. Climate change could also affect the international politics of energy production and consumption. Oil- and natural gas–exporting countries, especially Russia, could gain geopolitical footing relative to fuel-importing countries, such as those in Europe. Any oil- or gas-importing nation with high energy intensity could suffer disproportionately, owing to the difficulty of switching to alternative fuel supplies. As is aptly underscored in this book, energy infrastructure could also become more vulnerable, both in the United States and globally. Hydroelectric power generation may be substantially affected by reduced glacial runoff or by upstream nations' diversion of rivers in some parts of the world. Nuclear power plant cooling may be limited by reduced water availability, particularly for plants built along inland waterways. Increased numbers and intensity of storms could interfere with long-distance electricity transmission, already heavily stressed in the United States and elsewhere. In a future world where climate change leads also to an increased likelihood of state failure, solving the problem of vulnerable energy infrastructure helps solve two problems: heightened risks of resource-related terrorism and the need to find alternative energy sources in the near term that mitigate further climate change impacts.

5. More Nuclear Energy

Climate change may well lead to a global renaissance in nuclear energy generation, driven partly by the expectation that increased use of nuclear power plants will reduce the use of carbon-emitting fossil fuels. Such a development could worsen problems of nuclear safety and proliferation. According to current plans, many developing countries will begin operating their own commercial nuclear reactors during the next few decades. Several of these major nuclear projects are scattered across the Middle East in states with deep political uncertainties. This would increase the total number of nuclear reactors around the world, including those under the control of nations that may lack the experience to safely conduct these operations. The threat of global climate change also provides governments interested in acquiring nuclear weapons yet another justification to pursue nuclear-related research and nuclear technologies. For example, the oil-rich countries of the Middle East are among the largest emitters of greenhouse gases per capita; these nations could reasonably claim a need for nuclear power to help desalinate water or cut greenhouse gas emissions.

6. Social and Political Change

Severe or sudden climate change presents a profound challenge to existing social and political organizations in countries rich and poor. International cooperation might increase as people rally to save human civilization, or individuals and groups might become preoccupied with promoting their own survival. Under enormous stress brought on by climate change, the United Nations and other existing international institutions will have great difficulty managing the full range of adverse consequences. The implications of new international alignments driven by environmental factors are uncertain, but the complex and inherently divisive nature of climate change—because of the power and influence of the interested industries and energy players involved—means that forging collective responses will be exceedingly difficult.

7. Challenge to Existing Governments

Climate change could have deep implications for the very effectiveness and viability of existing governments. Political authorities unable to manage climate-induced challenges might well lose necessary public support. National leaders professing authoritarian ideologies could become more attractive if liberal democratic systems fail to marshal sufficient political will to manage the climate challenge. In some instances people might resort to violent means to remove existing governments, especially when opportunities to change leaders through elections are circumscribed. A succession of failed elected leaders might also trigger a rise in politics by other means as well. In a few places people might turn to nonstate actors, including religious movements or terrorist groups, for comfort or to effect more dramatic change. Moreover, under conditions of severe global climate change, environmental factors may push already failed states deeper into the abyss of ungovernability, while driving other states toward the brink.

8. World Political Disequilibrium

Given that problems ensuing from climate change likely will affect states very differently and some countries will respond more effectively than others, climate change has the potential to affect relations among sovereign states. Certain nations could require increased public health assistance in the face of urgent domestic needs; others could experience limits on their exports because of unanticipated changes in the global trading regime due to climate-related effects; others could become more vulnerable to foreign predation if a stronger neighbor suddenly desired their land or resources. Over the long term, the very divergent regional effects of climate change could affect the evolving global distribution of power with unpredictable consequences for international security.

9. Role of China

The economic and political decisions made by this generation of Chinese leaders will have a decisive effect on our global future. China is becoming the primary driver of global climate change, now emitting more carbon dioxide in aggregate (though not per capita) than any other nation.[1] A recent *New York Times* editorial denounced China and the United States for establishing an "alliance of denial" in which the two countries "are using each other's inaction as an excuse to do nothing."[2] Many members of the international community are calling on Beijing to adopt more rigorous policies to limit the growth of China's carbon emissions to reflect the country's status as an emerging global stakeholder sharing the burdens of world leadership. Some of these appeals have been less than effective, for China's reasoning that the United States is not showing itself to be serious still holds validity. According to the World Bank, sixteen of the world's twenty most polluted cities are in China—the air is so polluted that it causes 400,000 premature deaths every year.[3] Not surprisingly, China's own population is emerging as an important voice inside the country arguing for more responsive environmental policies.

10. Consequences for the United States

Americans are unlikely to escape the plausible adverse consequences of global climate change. If climate change results in a very substantial rise in sea levels, it could well cause widespread damage to New York, California, and other core industrial and agricultural regions of the United States, leading to a massive depopulation of U.S. coastal regions. In the immediate aftermath of any natural disaster, whether caused by climate change or other factors, the international community will look to the United States, with its unique world role and response assets (including those in the U.S. military), to assume a leading role in organizing the relief operation. How or whether the United States is able to perform this role effectively will contribute considerably to the perceptions of Americans in many countries. The new politics of global climate change will not obviate the need for U.S. policymakers, like their colleagues elsewhere, to continue weighing trade-offs among competing objectives and values, and managing climate change may not always emerge as the most important consideration. There is no question that climate change will mean fundamental shifts in how Americans see themselves and their role in the world. In all but the extreme scenario, in which most of the world is put in a fundamentally severe set of circumstances, the unique character of the American people, with their strong optimism and penchant for the practical, will be a major asset.

Conclusion

In the course of writing this report we arrived upon inescapable, overriding conclusions. In the coming decade, the current and the next generation of American foreign policy and national security practitioners will face an ominous set of challenges. These include reversing the decline in America's global standing, rebuilding the nation's armed forces, finding a responsible way out from Iraq while maintaining American influence in the wider region, persevering in Afghanistan, working toward greater energy security, reconceptualizing the struggle against violent extremists, restoring public trust in all manner of government functions, preparing to cope with either naturally occurring or man-made pathogens, and quelling the fear of new and unknown challenges on the horizon that threatens to cripple our foreign policy—to name just a few.

Regrettably, to this already daunting list we absolutely must add dealing responsibly with global climate change. Our group found that, if not dealt with proactively, climate change may come to represent a foreign policy and national security challenge as great as or a greater than any other problem on this list. Several participants came to believe that climate change would come to shape this and the next generation's views of the world and of what Americans' role in the world is, as the cold war shaped the last three generations of Americans. And, almost certainly, overarching global climate change will complicate many of these other issues.

Although all those who collaborated in this study completed the process with a profound sense of urgency, we also collectively are encouraged that there is still time for the United States and the international community to plan an effective response to prevent, mitigate, and, where possible, adapt to global climate change. We hope this study will help in that endeavor.

Indeed, the overall experience of these working groups helped underscore how much needs to be done on a sustained basis in this emerging field of exploration. It is hoped that this study will help illuminate how security concerns might manifest themselves in a future warming—and worrisome—world. Moving forward, the United States and other nations must chart a new path, for we already face a very real climatic cataclysm ahead.

Notes

1. Netherlands Environmental Assessment Agency, "China Now No. 1 in CO2 Emissions; USA in Second Position," press release, June 19, 2007 (www.mnp.nl/en/service/pressreleases).

2. "Warming and Global Security," editorial, *New York Times,* April 20, 2007.

3. "Environmental Challenges of Development in the East Asia and Pacific Region" (Washington: World Bank, 2006), p. 2.

About the Contributors

Sharon Burke is a senior fellow and director of the energy security project at the Center for a New American Security (CNAS).

Kurt M. Campbell is CEO and cofounder of the Center for a New American Security and is a former deputy assistant secretary of defense for Asia and the Pacific.

Leon Fuerth is a research professor of international affairs at George Washington University; he was national security adviser to Vice President Al Gore.

Jay Gulledge is the senior scientist and program manager for science and impacts at the Pew Center on Global Climate Change.

Alexander T. J. Lennon is the editor-in-chief of the *Washington Quarterly* and a fellow in the international security program at the Center for Strategic and International Studies.

J. R. McNeill is a professor of history at Georgetown University.

Peter Ogden is senior national security analyst at the Center for American Progress.

Christine Parthemore is a research assistant for the energy security project at the Center for a New American Security.

John Podesta is president and CEO of the Center for American Progress and was chief of staff for President Bill Clinton.

Julianne Smith is the director of the Europe program and the Initiative for Renewed Transatlantic Partnership at the Center for Strategic and International Studies.

Richard Weitz is a senior fellow and director of program management at the Hudson Institute.

R. James Woolsey is a venture partner at VantagePoint Venture Partners and is a former director of the CIA.

Index